DISAPPEARING WINDMILLS

*A Waist-High View
of the
World's Hot Spots*

Kate Christie Zee, M.D.

VANTAGE PRESS
New York

FIRST EDITION

All rights reserved, including the right of
reproduction in whole or in part in any form

Copyright © 1999 by Kate Christie Zee, M.D.

Published by Vantage Press, Inc.
516 West 34th Street, New York, New York 10001

Manufactured in the United States of America
ISBN: 0-533-12749-1

Library of Congress Catalog Card No.: 98-90268

0 9 8 7 6 5 4 3 2 1

Fortune disposes our affairs better than we ourselves could have desired. Look yonder, friend Sancho Panza, where you may discover . . . monstrous giants . . . with whose spoils we will begin to enrich ourselves.

—Don Quixote

Contents

1. The End That Turned Out to Be a Beginning — 1
2. Disappearing in China — 8
3. Excluded, Included, and In Between in India — 26
4. Krishna's Boots: Nepal's Magic — 44
5. Gaining a Sense of Proportion in Russia — 60
6. And Now for Something Completely Different: The One Journey I Would Rather Not Have Taken — 80
7. Facing up to Fear on the Amazon — 85
8. A Beggar's Eyes: Cambodia's Lesson — 103
9. Dancing to a Different Drummer in Vietnam — 124
10. To Control One's Destiny: India Revisited — 136
11. Vulnerability and Power in Tanzania: Balancing an Important Equation — 147
12. The Return of the Light in India: A Journey of Acceptance — 161

Epilogue — 181
Acknowledgments — 185

1

The End That Turned Out to Be a Beginning

"Make way! Make way!" shouted the paramedics as they ran full-speed with a moaning man on their stretcher toward my biggest trauma room. The hot, steamy air blowing in through the wide-open doors of the emergency room in their wake pushed softly on my back, encouraging me to swiftness as I rushed after them. I hoped I wouldn't trip.

"Not now," I prayed. "Not now." I had been falling a lot recently, but this was definitely not a good time to crash down on the floor. I was the only doctor on duty and there was no time to lose getting to my patient.

On the man's upper torso, pale globbets of fat and red strips of muscle circled what remained of a shaft of ivory bone. Looking like a slab of raw steak laid flat on his sweaty chest, this small, bloody disc was all that remained of his left arm. Although blood, temporarily retained like water in a kinked garden hose, had not yet started spurting from the large, severed artery, our team raced to prevent the sudden gory waterfall that could take his life at any minute.

With his scrunched-up face and muscle-plumped body, the man looked like a frightened baby. Stifling a quick impulse to rush into reassurances and stroke his undamaged shoulder, I concentrated instead on what I had to do to keep him alive. Newly facing a life–altering change myself, I now knew that in the face of such a disaster, the usual quick, easy litany of reassurance—the consoling mumbo jumbo that had been my stock in trade—would be meaningless.

As the paramedics held the man upright, I listened quickly to his heart and lungs, then looked for wounds on his back. Fortunately, except for the shallow, clotting gouge where his arm had once been, the rest of him was normal.

One by one, I crossed the usual tasks off my mental list as my two nurses and I did them—physical exam, placement of intravenous line, catheterization for urinary drainage, and electrocardiogram—with the knowledge that the rest of the team would be arriving very soon.

By the time I finished threading a special fluid-delivering catheter straight into the man's heart, the troops had all arrived: X-ray and laboratory technicians, nursing supervisor, and finally a trauma surgeon, hair still wet from a shower, who had rushed in from home. We were in a small community hospital that could not afford to keep doctors of all specialties on the premises at all times. But as we joined together in familiar routines, we made up for our lack of numbers by merging as components of a well-organized machine. Yet I felt oddly off-balance, as if I were some slightly dislocated cog. Even worse, the others seemed aware of it.

Not as intent on the patient as usual, they flicked glances toward me and away. I knew the look. It was the same I would give a stranger who inadvertently wandered in on a Code Blue while I was trying to start a stopped heart: a quick assessing glare for someone unknown, in the way, and possibly dangerous. I had once been smoothly integrated with this E.R. staff. They had been my comrades-in-arms through countless situations—dangerous, exasperating, happy, and sad. But now, to them, I had become an unknown factor, my presence personally threatening, no matter how hard they tried to hide it.

As I stepped quickly out of the way of the clanking behemoth that was the portable X-ray machine, I staggered slightly. I hoped no one would notice. But, of course, they all did.

I had not intended to tell them of my diagnosis with multiple sclerosis (MS) until I had time to get a little bit used to it myself. Confidentiality of medical information is, however, a luxury known only to laymen, if it even exists at all. Passing like wildfire through the medical neighborhood, the unwelcome news had leapt and spread to everyone I knew, and even many I did not. Conversations stopped when I approached chatting nurses on their break; keenly assessing eyes turned covertly on me as I worked.

Although confronted daily with contagion—infected blood, foul pus, germ-laden coughs—the nurses seemed unafraid of their contact with the sick. But now, to my surprise, they were keeping as careful a distance from me as they might from a leper. It was as if I had suddenly become a source of easily transmittable disease—my new vulnerability to illness transferable—if one got too close. With a potentially disastrous condition, I was the living example of what we all feared . . . and much too close for comfort.

When the trauma surgeon no longer needed my help, I left the room to see how many patients had accumulated in the waiting area while we all worked behind closed doors. As I walked into the hallway, I glanced back at the bloodied man staring with unseeing eyes at the ceiling. Amidst the crowd of people in surgical green, he looked very alone. And, in truth, he was. I didn't even know his name.

He didn't know or care anything about me, the other doctors, the nurses, or the hospital. All he knew was he no longer had his arm. Maybe he was trying to figure out how he was going to live; maybe he didn't want to. Facing a similar dilemma myself, I knew the feeling.

It had only been a couple of months since a neurologist, face carefully blank and gaze directed at his firmly clasped hands, had bluntly named my illness. There had been no gentle build-up to the news; no attempt to soften the blow. He knew I knew. I didn't question the diagnosis. Together we had watched my exaggerated reflexes when he struck my tendons with his rubber hammer; had seen my toes going up instead of down when my foot was stroked. We both knew that I had reported the prick of a piercing pin as merely gentle pressure and that my eyeballs flickered uncontrollably when I gave his moving finger a sidelong glance. With each event my heart sank.

I had come to see him complaining only of dizziness and of a tendency to trip on flat surfaces which, when I looked back at them, were entirely smooth. I had come to him with a hope that maybe he could discover a benign diagnosis. But when I saw these disastrous signs, I could no longer maintain any illusion that I had a simple, curable disease. My brain and spinal cord—the intricate definers of life as we know it—were no longer functioning properly.

Commands from my brain were now erratically transmitted through the nerves that have the job of relaying instructions to the rest of the body, and my body could no longer send reports back to the brain with reliability. This was because the electrical transmission of messages along the connections of the nervous system, the neurons, had been disrupted by patchy stripping of their insulation, the myelin sheath.

A simpler way of looking at it was that contact between command headquarters and the troops had become unreliable because insulation had been stripped from the wires conducting electricity through all their communication devices. Any attempt to relay information could be delayed, garbled, disrupted, or blocked entirely.

Such disruptions of my neuronal communications occurred

randomly and without warning. They could occur any time or any place. Neurological problems lasted an unpredictable length of time, and sometimes became permanent. A foot became paralyzed; an arm went numb; speech slurred; a bladder wouldn't empty when full. Lacking central control, reflexes reacted out of proportion to a stimulus, muscles spasmed periodically with no apparent reason. My body sometimes felt like a war zone, with the troops in mutiny.

There was no real treatment, no first aid battalion to patch up any injuries. Any known interventions were only Band-Aids tossed in the general direction of the troops. Even worse, there was no cease-fire, in the foreseeable future. The skirmishes would go on and on. I knew all this at the time of diagnosis, and found it hard to hope that I could ever escape the battlefield. There was no known cure.

I almost wished that, if disaster had to happen, I had instead suddenly become an amputee or paraplegic. After one devastating event, my losses would be confined to known parameters and, with training, I could learn to muster into usefulness any able parts I had left. This was not the case with multiple sclerosis. No one could predict exactly what would happen—what specific problems I would develop, or when. The only certainty was that my illness was progressive and I faced a life of increasing disability. The world as I knew it suddenly seemed to wobble wildly in its orbit.

As tears sprang to my eyes, the neurologist had given me a weak smile and a brisk platitude, "Don't lose hope—there's always hope—you don't have to expect the worst." His run-together reassurances had rung hollow. He hadn't sounded sure. For a moment, I had hated him.

What the hell does he know? I thought. *And how dare he say it?* Hope for what? A normal life? The discovery of a cure? A quick end? But he had quickly retreated into the detached persona of his specialty and didn't stay in the room long enough for me to ask.

Any solace I might have gained from a leisurely discussion of my condition, a comfort I thought I deserved because of being a fellow doctor, he had been unable, or unwilling, to supply. Somehow, I understood his subtle message that I had lost my membership in the club, that fraternity of physicians I had worked so very hard to join. The doors to the frat house were already closing. Instead of being the doctor, I was now a patient. I was now the one on the outside looking in. Two months later, on the day after I treated the man with the blown-off arm, the doors closed entirely.

Called to a private meeting, I squirmed with embarrassment as the head of my professional group reprimanded me for easing out of duties I had been recently finding difficult. Oddly, he didn't comment on my newly discovered disability, nor did he mention any worries that—needing more time off than most—I might be a burden on the practice.

While I was trying to think of how to apologize and to ask for accommodations that might help me do better at my job, I suddenly heard him say, "I want you to submit your resignation."

Stunned into silence, I could say nothing. With my disability being treated like a dirty little secret, even as my mouth opened in disbelief I could find no voice with which to discuss solutions to any problems the disability might pose. This was 1982, eight years before the Americans with Disabilities Act was enacted. My world, already unsteady in its orbit, came to a complete halt.

I stayed home a month, crying and sleeping a lot. I guess that's what most people do when their world ends. Clutching the thought that no one had yet taken away my medical degree as a tiny shred of hope, I next wandered from job to job, sharing my knowledge with insurance companies and lawyers. Life began to look a little better, but not by much. Something was missing. I was only marking time, although at the time I didn't know it.

The beginnings of a new future, far better than the old, came sneaking in on the coattails of an old friend's suggestion. One evening, he handed me a brochure and said, "You've always loved to travel. There's a wheelchair tour leaving for China in November. Why don't you go on it?"

Spotting a picture on the cover of my worse nightmare—a wheelchair—I tossed the brochure on a table and said, "I'll look at it later." I meant to do nothing of the kind. But after he left, I picked it up, attracted by a picture of the Great Wall.

As I bent my head to read the itinerary, which hustled excitingly from Beijing to Hong Kong, I thought, "Well, I've certainly got the time. Why not do it?"

The decision was clinched by my remembering a T-shirt I had once seen which read: "I Climbed the Great Wall." I suddenly wanted one more than anything. Going to get it would serve as proof I could still be the adventurer I had once thought myself to be. Maybe life as I had known it wasn't completely over.

"I'll go," I said, as hope crept into my heart. There was no way to know that that split-second decision marked the start of a series of incredible adventures and a grand journey into wellness.

2

Disappearing in China

When I traveled to China, I hadn't expected to disappear. I don't think any traveler does at their journey's beginning. En route to Beijing, however, unheralded by anyone but me, my disappearance was already beginning, creeping up like old age on an unsuspecting adult in the prime of life.

A person's disappearance can come in many different forms. You can become a face in the crowd, the stranger nobody knows, or an adventurer away from home. Disappearing can be deliberately sought or thrust upon you unexpectedly. It can be as direct as joining the French Foreign Legion, or as indirect as finding yourself ignored. It can be as simple as seeing your health vanish when you awaken with the flu, as complex as seeing youth vanish when you notice wrinkles of age in your mirror.

We all disappear at one time or another. But until I used a wheelchair, I never expected the many ways in which one can disappear in it.

The first thing to go was a sense of personal autonomy. After the airline insisted I replace my newly rented wheelchair with a less comfortable one of theirs, it started to dwindle. It decreased even more when, deemed unable to pull open an emergency door if it became necessary, I was not allowed my favorite seat next to the exit. The worst moment, however, came on arrival in Beijing when I, a long-time solo traveler, couldn't even get out of the airport on my own.

As the only seated person in the crowded baggage claim area, I was an anomaly in the busy terminal as I waited for my companion. Hurrying travelers divided and flowed around me the way a river rushes around a rock in the center of its bed. No familiar signs announced an exit or the location of a taxi rank. No one spoke enough English for me to ask.

Already uneasy about traveling this way—with a wheelchair, with a companion, with a group, all new events to me—I fought off tears as I unexpectedly felt like a travel novice. Never before worried about such things, I now felt, as everyone ignored me, that something more than my autonomy was disappearing.

Not yet quite comprehending the extent of this unexpected event, I continually peered around the airport, hoping to spot our pre-arranged driver rushing toward me, or my wandering travel companion, T.C., returning. After what seemed a very long time, unsuccessful in his search for the driver, my friend came back alone clutching a small square of paper covered by a scribbled mass of Chinese characters. Looking like finely penned chicken-scratches, they were to serve, in lieu of an interpreter, as directions for a taxi driver.

During the surprisingly long and leisurely drive into the city, internal tensions began to ease. The wide, tree-lined streets near the airport were green and pretty under the gray, darkly polluted skies. Even more reassuring was the luxury of the hotel lobby we were taken to where what seemed like miles of glossy, rich brown marble glowed welcomingly in the light of giant, crystal chandeliers. At least a story high themselves, the chandeliers' radiance was magnified by the rebounding reflections of their light from the shining, polished ceiling high above. Perhaps I had not come too far out on a limb.

Upstairs, I was almost glad to discover, however, that the sterile perfection of the lobby did not extend to all the rooms. All that magnificence seemed far less intimidating when one noticed that the bathroom floor tiles fell an inch short of the base of the Western-style toilet. This discrepancy lent a sense of adventure to our stay.

A perfect lobby, reminiscent of an upbeat mausoleum, was to be found in every hotel along our modern caravan route. Yet, each one, like the one in Beijing, had some slight imperfection in the upstairs rooms: a shower floor without a drain, a small round hole in the ceiling above the bed, a window made drafty by a frame unmatched to masonry.

Such architectural anomalies were most likely a manifestation of the growing pains of this rapidly developing country. Much of central Beijing was hidden behind high bamboo scaffolding that ascended the many new hotels under construction. Swaying slightly even when there was barely a breeze, these fragile-appearing supports seemed to desperately cling with tiny, knobby fists to the rough cement walls. The speed with which the hotels were being erected may have been a tribute to Chinese engineering, but perhaps the timetable was too hurried to allow full attention to all of the small finishing details.

The very subtle imperfections in these smaller details became, for me, a symbol of modern China—magnificent exteriors for public viewing but with always a minor flaw hidden privately behind closed doors. Perhaps I was a symbol in reverse, with a body slightly flawed in the public view but housing a magnificent interior for private inspection.

A few hours after our arrival, our assigned driver/interpreter appeared. Broad-faced and plump, he appeared unsurprised that we had found our own way but repeatedly explained, "I looked for you all over. I couldn't find you. Nobody saw you."

After ascertaining that we had all been at the airport at the same time, I couldn't understand how he could have missed the definite ripple my waist-high presence created in the flow of hurrying travelers or the towering figure of T.C., who at one point had jumped atop a baggage carousel to scan the crowd. Deciding that perhaps it was because neither of us had been at usual sight level, I missed the significance of this first clue about the odd cloak of invisibility that fell upon me now that I entered a wheelchair. Odder yet, the same seemed to apply to my companion.

This new invisibility became difficult to ignore when T.C. and I went out to watch the city wake up the next morning. We had arrived five days ahead of the group to have time to get acclimated. As T.C. slowly pushed me along almost empty gray streets, we passed like ghosts among the few people stirring. Fragrant steam rose from cauldrons of breakfast soup nestling among hot coals in huge, rusty, iron drums. We passed very close to a hollow-cheeked man who warmed his hands in the rising vapor. Shivering as he crouched over the red-gold glow of the burning embers, he did not spare us a glance. Neither did the old men in the parks posturing gracefully as they practiced Tai Chi, nor the young men on the curbs brushing their teeth by chewing on thin wooden twigs. No one acknowledged our passing. We went unobserved and unremarked upon in the strange foggy netherworld of that gray and chilly dawn.

Later, in a Chinese medicine hospital, being unnoted was slightly less remarkable. T.C. had injured his shoulder when he yanked my bulky wheelchair off the baggage carousel, and was quickly spirited off by a young, English-speaking orthopedist for examination. I sat in a corridor waiting for a different specialist, one who could tell me if Chinese medicine could offer any effective treatments for MS. Although I was not expecting such a miracle, it was worth a try.

Familiar with the invisibility of hospital patients in wheelchairs, I was not surprised that dozens of medical women, recognizable by long, white lab coats and white, drum-shaped caps, hurried by without acknowledging my presence. Finally, a female doctor looked directly at me, then split off from a chattering crowd of her peers. Her smile made her cheeks protrude like ruddy, wizened apples in her dry, wrinkled face as she motioned me to a quiet corner. Her twinkling eyes were kind as she introduced herself and shook hands, then started gently patting my spine.

"I will translate for you," said the driver, beaming with confidence.

I told him to tell the doctor that I had MS and would like to know if she could help. The driver nodded encouragingly and then translated all I had said into only two words. Smiling at me, the doctor patted me more firmly on the back, then delivered a lengthy speech. As her voice melodically rose and fell, the driver's nods were enthusiastic.

"She will help you. She will admit you for back surgery," he said a couple of minutes later when she finished.

She had given a long speech. I waited for more information. None came. I thought perhaps I should fill in the void. "Tell her I can walk some," I told the driver.

As he started to translate, I stood, then hobbled a few yards down the hall and back. The tiny doctor looked startled. She had apparently assumed I could not walk at all. Lacking adequate communication, any hope of mutual understanding had disappeared behind a barrier of false assumptions based on my new means of locomotion. And she was a doctor! The drafty hospital hall seemed a bit colder as I unwillingly recognized this as an omen of what might be ahead if I ever needed the full-time use of a wheelchair.

After a brief discussion with the doctor, the driver nodded uncertainly, then explained with some embarrassment that they did not know of my disease. I turned down suggestions for acupuncture and for five hundred dollars worth of herbs (guaranteed to work after I used them for at least a year). The doctor soon politely excused herself and left. I had used up enough of her time.

After the driver went to look for T.C., I was alone in the hallway. With no one around to watch, I practiced small tight circles and figure-eights. Maneuvering the wheelchair was far easier than I had expected and soon, emboldened, I decided to sneak into a nearby pharmacy. On entering, it was as if I had crossed the threshold into another century. This was no sleek, modern pharmacy with shining bottles of potions and pills arranged on bright metal shelves. Instead, narrow cluttered paths wound in a Minotaur-like maze around jars of moss, baskets of bark, and bundles of dried branches.

Stifling an urge to sneeze, I carefully maneuvered through the two large, deserted rooms. Suddenly, fearing a penalty for being where I didn't belong in a Communist country, I was in a hurry to get out before someone came in and discovered me. Surely in here I would be noticeable. I got out just in time. In less than a minute, amidst a crowd of doctors, T.C., clutching a tin of Tiger Balm, came with the driver to collect me.

After finding, to our surprise, that we would be allowed to explore the city unescorted, we set out to explore Beijing alone. In a city with hundreds of thousands of people on wheels, my method of locomotion was not particularly remarkable. Traveling in the wide, smooth bicycle paths between street and sidewalk, we merged smoothly with hundreds of people riding bicycles, pedicabs, and ox-drawn carts, or pulling long, flat-bed carriers filled with produce. A few disabled Chinese in government-supplied, red motorized carts passed us rapidly as they drove along the street.

Produce pullers with wide, toothless smiles occasionally pointed at me and mimed that I could ride in more comfort if I rode with them. Amidst these friendly workmen, my new invisibility had dropped like a discarded cloak.

T.C. jokingly pantomimed to one elderly man that we would race. Promptly gripping the handles of his cart more tightly, he gave a smiling nod. Assuming they were teasing, I missed the start signal while grinning back at the beaming old man.

Suddenly my wheels were accelerated to high speed as everyone nearby called encouragement and began to run. Squealing, I clutched the wheelchair's padded arms and held on tight. Unnerved by moving so fast over unfamiliar territory, I yelled for T.C. to fall back and let the old man win. Both men took my shouts as further encouragement. Faster and faster we went on the bumpy road. The men seemed to enjoy the game—and my protests.

Suddenly I found myself freewheeling, pushed ahead of T.C. like a fast-moving shuffleboard disc. Frantically scrabbling for the brakes, I screamed and finally skidded to a halt. Furious at having been a game-piece rather than part of the game, I glowered while the men bid each other a laughing good-bye. But after being safely stationary for a few moments, I could admit that the race had been fun. Even better, I had reappeared in my own mind as an adventurer. I bent my head to hide my smile; I didn't want to give the men the satisfaction.

My new sense of self didn't last long, however. When I climbed the Great Wall the next day, I became the unwilling recipient of the most dehumanizing, soul-stealing stares I had ever encountered. At first, I didn't notice my audience as I was carried up by a trio of sweating men: T.C., our driver, and a Chinese soldier who had rushed over to help. Terrified at being dropped off of the tall unevenly carved steps at the foot of the watchtower we were climbing, I was too busy concentrating on the slipping hands of those lifting my wheelchair.

Of differing heights, the men grunted as they struggled to hold the wheelchair level. It swayed and teetered precariously as they tried to match their strides. Shorter than the others, the soldier adjusted his grip several times, coming perilously close to the lever that detached the right leg-piece from the frame. Gritting my teeth, I worked to repress visions of the damage I would sustain if the leg-rest suddenly flew off. When I was finally deposited on the top, it took several minutes before I could pry my fingers away from a death grip on the armrests.

The width of the flat expanse on which I sat bore mute testimony to the thickness the Chinese had thought necessary to protect themselves from successive waves of Mongol invaders many centuries ago. Now, it provides a broad trail for tourists to hike up and down as the Wall sinuously follows the conformation of the rolling hills.

Realizing that the slope further towards the top of the next hill was too steep for him to maintain adequate control of my velocity, T.C. hiked alone toward the next guard tower with a promise to photograph the view I wouldn't be able to see. Soon he disappeared into the mass of people who, from afar, made the twisting Great Wall look like a writhing, multicolored dragon.

The Chinese hills stretched as far as I could see. As I strained my eyes to follow the winding, curved outline of the Wall, I also thought I caught a glimpse of the far-distant Silk Road, the ancient trade route to the Middle East. Then, as covertly as possible, I studied the seemingly infinite variety of Asian tourists passing nearer by. I tried to disguise the fact that I was watching them, but soon discovered I needn't have been so polite.

Asian tourists stared at me upon approach, while passing, and even after they had gone several yards beyond. So great was their concentration on me, they inadvertently ran into others in their path who, also staring at me, were unable to get out of the way. Even if I stared back, looking directly into their eyes, no one looked away. Oblivious to Western ideas of courtesy, they watched me until I was out of their sight. No one said a word. No one smiled.

Unblinking and unwavering, several hundred pairs of obsidian eyes watched me as I sat. No trace of my identity was reflected back from their expressionless depths. I felt as isolated as an object on display in a Bell jar as all traces of my humanity, even my soul itself, threatened to disappear into the vacuum of that utterly fathomless stare.

Even though I didn't relish a repeat of the uneasiness I had felt while being carried up the Wall, I was relieved when it was time to go back down, especially when T.C. agreed to help me limp down the steps. Under the concerned eyes of our driver, I concentrated on my feet—not daring to look back to see if I was still being watched—and joked that I'd be the first tourist that ever climbed only *down* the Wall.

As we were being driven back to Beijing, I asked if we could make a side-trip to the Silk Road, but our driver informed me with a tolerant smile that it was over two hundred miles away. My vision of it in the distance had been only a mirage. My mind's eye had peered over the hills into infinity, going further in mere seconds than even wheels traveling at the speed of sound. Only if able to travel through time at the speed of light could anyone have taken me to where I had seen dim caravans passing slowly by, accompanied by the sound of tinkling camel bells under the moonlight.

* * *

When I returned to the Wall five days later, now in the company of six more people in wheelchairs, the stare was magnified exponentially but was completely changed in quality. When confronted by a whole group of wheelchair travelers, observers showed a disconcerting multitude of feelings. Concern, worry, curiosity, pity, and other emotions too complex to decipher all competed for expression on the faces of those nearby. The air vibrated with a hundred conversations as people called their friends and children to join them, creating a mob scene frightening in its magnitude and intensity.

Within minutes, whistle-blowing police hurried over to shout rough inquiries. With harsh-toned words and fierce scowls, they shook thick, black truncheons to quickly disperse the crowd. Feeling vaguely guilty, I waited to be reprimanded for my part in inciting a riot. But, after a quick glance, the officers ignored us while looking grimly about to make sure the crowd was staying away.

Things were almost back to normal when a boisterous amputee with our group decided to clear a path for herself by operating the noisemaker she had bolted to one of her armrests. Obscenely loud imitations of sirens, both British and American, stridently alternated with the sounds of a noisy assortment of bells and whistles.

A quick fear arose in the faces of the Chinese tourists, who scattered instantly. When they glanced back over their shoulders, the amputee motioned them closer with a smile. When she demonstrated the origin of the alarming sounds, their tentative smiles and shaky laughter were akin to those one might show after discovering his house had not burned down after all, or his child was not really dead under the truck in the street. At that moment, I would have welcomed any way to truly disappear.

After the ride back to downtown Beijing, the offensive noisemaker turned up missing, but no one voiced condolences. I vowed never to use such a device. But sometimes polite silence is not always the best policy.

The next day, sitting slightly behind the group in the Forbidden City, a huge compound once the closely guarded home of emperors, I was enjoying the sight of my fellow travelers in front of the series of archways we were about to enter. The symmetrical openings diminished in size as they progressed into the distance, and within the center of each could be seen a building. As each of these also diminished in size, they drew the eye in like the drawings used to teach children about infinity. In old China, this sight had meant death to anyone who had entered the city unbidden.

Suddenly, my heart leapt when, with no warning, I found myself being pulled backward and then pushed to the front alongside my companions. Letting out an involuntary squeak, I could have been no more startled if one of the ancient guardians of the emperor had suddenly materialized and started to drag me off to my death. Peering up and behind me, I saw that the man silently moving me was someone else's companion.

Struggling for composure, I tried to keep my voice soft enough to not disturb the group as I exclaimed, "What are you doing?"

Cheerfully smiling, he said, "I thought you'd like a better view."

"But I was happy where I was," I murmured.

Looking hurt, he asked, "Do you want to be moved back?" I nodded. His face set in lines of annoyance and resentfulness before he returned me to my place. The only similar reaction I could recall ever receiving was in earlier years when I gave my own order to a waiter in the presence of a date who preferred to be a "gentleman" and order it for me. It appeared that I now had unwittingly committed a similar kind of faux pas, and now, just as then, I felt awkward and uncomfortable in dealing with the situation.

Chinese hotel clerks were even better at this mission of setting me in undesired motion. Politely and silently, they would routinely move me like a piece of castored furniture away from the front desk while T.C. was checking us in. Equally silent and polite at first, I would allow them to place me facing a plain blank wall instead of the elegant lobby. Studying these walls of marble, wood-paneling, or plaster without even a painting to break the monotony, I puzzled over why I had been placed facing them. Giving the clerks the benefit of the doubt, I assumed at first that they were merely getting me out of the way of the crowds checking in or out at the desk. But I finally got the real message when we came to a lobby decorated with tropical greenery. After being moved toward a pleasant seating group of sofa, coffee table, and two chairs, I found myself being placed amidst, not alongside, the ferns to the side of the sofa. I was wanted not only out of the way but also, as much as possible, out of sight. No matter how firmly I protested, no matter how nicely I smiled a plea to remain by my companion's side, I was still moved. My only recourse was to wait until their backs were turned, then make my own move to wherever I wanted to be.

Such self-determination about placement was not always an option, such as when our group was taken to a Friendship Store, a government store purveying tourist trinkets and Western goods. On an upper floor, the store could only be reached by climbing a steep set of stairs. While the able bodied ran up for souvenirs, the disabled were left to wheel around at ground level and see whatever they could.

As I waited, I asked a local guide accompanying the group how they thought a group in wheelchairs could get up to that particular store.

Shrugging, he said, "I hear you go up Great Wall. We didn't think this any problem."

Wheelchairs may have been viewed by the Chinese as simply a tourist oddity, receiving no more attention than a bag bristling with cameras and spare lenses. Perhaps no one understood that the wheeled members of our group might want to go somewhere unassisted. Whatever the reason, it was my first experience of the frustration of being stuck on ground level when wishing to join a group of friends convened at a place accessible only to the independently mobile.

Sometimes, however, we all were taken to places accessible only to a select few, like the day we all went to the circus in Shanghai. When the show began, I was in touching distance of the panda bear as he bicycled slowly by. Wiping away happy tears, I watched as, only yards away, tumblers tumbled and jugglers juggled in glittering succession while the flying men swung directly above. All entered the ring from ramps between the bleacher-like seats, one to our right and one to our left. It appeared that our first-row seats were the best to be had in the almost completely round theater that encircled a single large circus ring.

The setting was intimate—almost too intimate—for the grand finale where all the bows were directed away from us. Puzzled, I looked at the compact, tightly muscled derrieres of the performers as they were presented only a few feet away. Finally, I realized what was wrong. In a round theater, they had made the best possible attempt to place us backstage, as out of sight as possible.

When our group later went to a restaurant, however, even the best attempts by the Chinese to position us out of sight were doomed to failure. After arrival, I happened to be the first to exit our unmodified bus. With seven collapsible wheelchairs loaded end-to-end filling the aisle, it would take at least forty-five minutes more for our American escort and two Chinese guides to off-load everybody and their wheels. Unwilling to face the stares while the rest of the group disembarked, I virtually hurled myself into my wheelchair when it was brought out and the usual crowd began to form.

After I raced through the visual gauntlet into the safety of the restaurant, consternation spread over the face of the elderly man at the registration desk. Lunch had been set up on the second floor—and there was no elevator.

Several serious men in dark business suits hurried over for an increasingly excited discussion as I explained that there were six other people in wheelchairs yet to come. Soon our American escort was brought in to join the worried-looking group.

Finally, with handshakes and smiles, the group split apart. The elderly man, still at his desk, snapped his fingers at several slender, white-jacketed youths who hovered in the doorways of the entrance hall. Running, they soon returned with a large round table and placed it in the very center of the room, directly in front of the front door. It appeared there was no accessible place where we could have lunch in a less visible manner.

Within a couple of minutes, seven places were set. No chairs were brought. It was obvious who was to sit downstairs and who up. I didn't want to sit in my wheelchair. It was not only because it was much less comfortable than sitting in a chair designed to be the right height for a particular table. I was also reluctant to sit in my newly assigned place because I had never before been in such exclusive contact with a large group of people in wheelchairs. Like many of my able countrymen, I unfortunately felt a little uncomfortable around them.

Longing to leave, I wanted to join the companions and assure myself that I was still normal. It didn't appear to be an option. I was stuck.

Sighing, I sat back to get to know the group. It went much better than I had expected. During the family-style meal at a large, round table, we spun a round disk in its center like a giant lazy-Susan to bring the many serving dishes close to each person's place. With a little cooperation, nothing went spinning out of anyone's reach and soon everyone settled down to the elaborately presented multi-course meal.

As it will among most strangers, conversation first stuttered, then settled smoothly into familiar social rhythms. I was surprised at the ordinariness of the event, and embarrassed at my own previous prejudices. Even though I was having a good time, however, I was still afraid to identify with this new peer group too closely. As a woman, I had already faced a battle to be recognized as an individual—fighting assessment based on gender instead of actual talents—and had no desire to gird up for any similar struggle precipitated by disability. It would take time for the fear to disappear—in slow, gradual increments—as the newness of disability wore off. I was now only taking my first baby steps toward emotional freedom.

* * *

As my visit to China continued, however, I increasingly realized that my worries about disappearance and loss of individuality were nothing compared to what faced Chinese females in their first few days of life. My worries about disappearing were figurative; theirs were literal, and many wouldn't get to take any baby steps at all.

Although girl-killing was denied in the official version of Chinese family-planning, in an almost empty restaurant in Beijing, an earnest young waiter told us it still went on: an unwanted girl baby simply disappeared. Plump-cheeked and chubby, bulky in padded jackets and pants, the babies of China were so beautiful it was hard to imagine that any of them could be unwanted. Yet, in a country which mandated only one child per family, parents strongly preferred a boy—a form of social security for their old age.

After first checking over his shoulder for unwanted listeners, the slender waiter quickly drew a finger across his throat to explain the fate of female babies. When his companions urgently warned him not to speak further, I shivered and gave silent thanks that my birth mother had chosen to give up her unwanted girl baby for adoption. Soon, after looking all around to make sure they were still unobserved, the young Chinese drifted quietly away from our table.

Even in a society with seemingly strong tendencies for making what was unwanted disappear, such cold pragmatism could still be overcome. There was yet a ray of hope for us all. This lesson was brought home in Nanjing, the southern capital of old China. In that graceful old city I found living proof that love is an emotion with more power than even that of an oppressive government.

As T.C. and I wandered about apart from the group, we noticed several tiny three-to-four-year-old girls shyly peering out from where they stood almost hidden behind the cinnabar red pillars of a Confucian temple. With dark sparkling eyes made hugely beautiful by make-up, they looked like elaborately painted porcelain dolls. Eye shadow, mascara and eye liner—nothing was spared. Tiny perfect lips reddened with bright lipstick parted prettily as they smiled. I waggled my fingers and smiled back, wondering at their appearance.

Enticed by several balloons that T.C. had brought with him, they came hesitantly closer. As T.C. slowly blew up the balloons and handed them out one by one—after discovering that no child would accept a balloon that was pre-inflated—they whispered to each other and giggled. Nearby, their watching mothers beamed with pride.

Once they all had balloons, the little girls ran off laughing. Later our local guide, a slender man barely out of his teens, explained the make-up. "Mother loves, even if only girl," he said, then shyly ducked his head.

Another man we met in Nanjing could not have been described as shy. As we wandered about, we became aware of him keeping pace with us at about fifteen feet away. If we stopped, he stopped. If we went on, he went on. He was giving me more than the customary stare. He was taking photos! It was with some trepidation that I noticed him rapidly closing in.

Bowing and smiling broadly, he informed us in almost unaccented English that he was a reporter for the Nanjing newspaper. After apologizing profusely for stopping us, he requested that I pose for him. Sure that his previous photos would be blurred by motion, he wanted a better picture for his article. It seemed that 1988 was the official "Year of the Tourist," and I was to be evidence that China was such a desirable tourist destination that even people in wheelchairs came.

Before reaching Nanjing, I had already grown to appreciate that I would not have been on this trip unless I had been disabled. I had come hating my need for a wheelchair, and had developed a strong dislike for the way I sometimes was treated while in it. But, after nearly a month in China, I was now concentrating more on the grand adventures I could have, instead of on the physical capabilities I lacked.

Thus it seemed the ultimate in irony that I should be chosen now for my few minutes of fame because I was in a wheelchair. Because of my means of locomotion, the reporter thought that, unlike most tourists, I was newsworthy. As a final assault on my ego, he informed me that he planned to draw a stirring analogy between me and the premier's disabled son. And I had thought he had just wanted a picture of a happy Western tourist.

I had consistently refused to let T.C. take my photo in the wheelchair, hoping to avoid later confrontation with a handicapped image of myself preserved forever in a faded scrapbook. I preferred pictures to be taken while I stood. A snapshot T.C. took, despite my protests, shows a weary traveler with a Mao cap, complete with tin medals and red star, tipped rakishly over one eye. Lounging in a beat-up old wheelchair, near the door of an equally beat-up bus, she is smiling with both resignation and happiness into the camera. It was the truer picture, and I'm proud of it now.

When I began the journey to China in a wheelchair, I hadn't expected to disappear—to have my identity and autonomy disallowed and unacknowledged—in so many ways. Even so, the wheelchair, instead of holding me back, brought me access to the far horizons of my dreams. In it I had even done more than many of those far abler than I. Although the trip was rugged, both physically and emotionally, when it was over I reappeared with a new sense of confidence and wellness.

Before I went home, I not only bought my "I Climbed the Great Wall" T-shirt, I also bought one that said in bright red letters, "I Survived China." The two of them pretty much said it all.

*   *   *

Two years after surviving China, a desire for more adventure sent out an irresistible clarion call. I had heard it all my life. Delighted by the knowledge gained in the Far East that I could still answer it, I soon plunged headlong into the colorful maelstrom of Indian society.

3

Excluded, Included, and In Between in India

As our hired car hurtled along the Delhi–Agra Highway toward India's famed Taj Mahal, I was blissfully unaware that we were driving on the wrong side of the road. During the long, hot journey, I had been caught up in daydreams about how a woman could have inspired such great love that she would be immortalized with such beauty.

Suddenly horns blared. Instantly jerked back to reality, I was facing the prospect of imminent disaster. Only halfway through a ten-car pass, we were speeding toward a head-on collision with a giant TaTa truck.

Looking like gnashing metal teeth, the truck's silver grill grew larger and larger until it appeared ready to bite through our windshield. The wall of cars to our right seemed as solid as if it were made of bricks and mortar. The truck roared closer and closer. Drivers held their horns down with a deafening, continuous blare, but no one gave way. Suddenly, the moment of truth was at hand. At the last possible second, as if directed by a choreographer, the truck ahead danced a few centimeters to our left, and the wall of cars danced a few centimeters to our right.

After squeezing through the narrow gap, our driver continued undaunted on the wrong side of the road. Oblivious to the ongoing peril, he pointed out a grassy slope almost covered with near-fluorescent red and yellow cottons, newly dyed and drying near the road. I spared only a quick glance for the colorful sight before I again fastened my eyes on the cars we were passing. I watched them intently, as if by staring hard enough I could create an opening that would allow us to return to safety. But the cars, moving at a stroller's leisurely pace, stubbornly remained bumper-to-bumper. Finally the cause of the blockade came into view.

At the head of the line, a plodding brahma bull slowly pulled a cart made heavy by a towering mound of hay. Dust kicked up by his hooves formed a small cloud around the ankles of the sun-darkened old man walking slowly behind him. Although he held a stick, the old man did not use it on his charge. When our driver abruptly cut back into the proper lane only a few feet ahead of the bull, the hump-backed animal did not appear to notice. Eyes front, tail switching lazily, he calmly maintained his steady forward pace. His master did not acknowledge our presence with even a glance. It did not seem to bother him that the modern world was passing him by. I wished I could remain as placid when, unable to move in an agile way, I worried that the life of a more mobile society might pass me by, leaving me behind as surely as the old man and his cart were left behind by those racing by in their cars.

Something as simple as not possessing a car was not what separated me from the social mainstream. Instead I was cut out by less tangible barriers—problems with access or the presumptions of others about my abilities. Unable to participate in many of the usual things that I had done when more agile, I was not intentionally excluded from my able-bodied society. But, often, I found myself side-lined, which feels about the same. I had not come to India expecting to explore the dimensions of being left out of the mainstream, nor did I expect to find solutions. However, bombarded by the sensory onslaught of that overpopulated country, I unexpectedly discovered that answers can come if you simply keep your eye on your goals.

The original goal of this journey had been to see the Taj Mahal. Blinding white and peaceful in its symmetry, this monument to love is so magnificent that it is one of the wonders of the modern world. But enjoyment of it can only be bittersweet. Although inspired by a man's great love for a woman, it was built because the object of that love had been lost.

As the subject of countless photographs, the Taj seemed at first a place I'd been before. Up close, however, as it thrust in on every sense, it seemed much less familiar. It was not pure white but instead looked like an iridescent rainbow as sunlight played off the hundreds of multicolored gems inlaid in the gleaming marble. At intervals, holy verses from the Koran traced in onyx were inscribed vertically on tall, gleaming columns. By some magical feat of perspective, they appeared the same size at the bottom as near the top, many meters above.

After discovering that the real tombs of Shah Jihan and his love were at the bottom of a steep flight of stairs, I chose to visit the imitation crypts which could be reached by climbing up an outside stairway. With an unstable leg, it was always easier to climb up than down. I was already halfway to the fake tombs before I figured out my mistake in logic. No matter which was my first choice, I'd still have to make one climb up and one climb down.

The steps were each nearly a foot tall, and there were well over a dozen of them. One slow step at a time, my good leg took the brunt of the climb, lifting me until my bad one could drag itself after. With anxious eyes, our hired driver let me grip his arm as he climbed slowly with me, while T.C. wrestled to bring the wheelchair up behind. Step, grunt, lift, rest . . . Step, grunt, lift, rest . . . As sweat dripped out of my eyes, the ascent seemed to take forever.

Suddenly, chattering like a flock of pigeons, about twenty

gray-veiled pilgrims appeared above. As they hurried past, the stream of women divided to the left and to the right, much as the truck and cars had divided around us only a few hours before. With eyes briefly cast down and heads gently inclined, they pressed their palms together and welcomed me gently in Hindi. "*Namaste,*" they said, almost whispering. "*Namaste . . . Namaste . . .*" The word echoed softly among the group.

"*Namaste,*" I said, inclining my head in turn. But, fearful of tumbling down the steps, I did not release my hold on the guide to press my palms together and complete the greeting.

Once inside the dim, cool tomb, I dropped gratefully into my wheelchair. While I gazed in awe at the jewel-studded sarcophagus, T.C. left to investigate the real tombs down below. But I was not alone. A few remaining pilgrims with reverently bent heads slowly rubbed their hands on the marble box, then on their foreheads. Their lips moved silently as if in prayer. Although this was an empty mausoleum, not a temple, the hushed atmosphere was one of worship. I wondered if they were praying for love— love as great as that which had inspired this incredible memorial. I decided to pray, too, although I barely dared hope for the kind of love that had inspired the Shah.

Many American women seem to worry about finding love, fearing as much rejection by others as they place on themselves because of perceived physical defects. They feel too fat, too thin, too tall, or too short; they think their hair is too curly or straight, too thick or thin, or the wrong color entirely. They are not satisfied with the shape of their noses, the size of their breasts, or the color of their eyes. Billions of dollars are spent by those who wish to create a desired physical appearance, to accent the good points and diminish the bad. I was no exception.

But no matter how much I could spend, there were no magical powders or potions that could make a recalcitrant body part work or create energy when fatigued muscles declared a time-out. Unable to improve or hide these physical defects, and having found no one yet who could ignore them, I felt that love and marriage would probably leave me on the sidelines. Yet hope dies hard.

Even the Mumtaz Mahal, the beloved wife of the Shah, could not have been all that ideal. She could not have been, by any stretch of imagination, a beauty of Hollywood fancy with slim figure, diaphanous veils, and a jewel in her navel. She had lived in an age without deodorant and even without good soap. After thirteen children, her body was most likely quite unappealing before it broadened and sagged even more during her last pregnancy, with a fourteenth baby, the one whose birth had killed her. But she had brought to her Shah such love and inspiration that she had been recorded by history as beautiful. Perhaps, if one had known her, she was.

I waited for the moment when I was completely alone. Finally, with a bow toward the tomb, the last woman drifted out. Alert for any sound from the doorway, I stealthily rolled closer. I glanced quickly around and then, in imitation of what I had seen, rubbed the sarcophagus and then my forehead, as I prayed a vaguely Christian prayer, asking to find such love. The peaceful mood felt as if it would last forever, but it lasted only as long as it took to leave the Taj by the exit gate.

Just outside the gate, some of India's hordes of crippled beggars were amassed. The politically correct description, "something impaired," would have been ludicrous. "Impaired" could not begin to describe the immensity of their situation. Men and women with heart-wrenchingly obvious deformities held out their hands, if they still had them to hold out. The rheumy, inflamed eyelids of one old man opened around white, sightless eye tissue. An eye-slashing scar distorted another's face, fixing his lips in a permanent sneer. Yet another's face was frozen into a rigid mask by scars of untreated burns. Oozing sores plagued another and yet another. Some had grotesquely twisted legs; some, no legs at all.

There were no wheelchairs, seeing-eye dogs, helper dogs, or crutches; no prostheses, eye patches, or even simple bandages. Even though my wheelchair was a dirty tan rental, for them it would be an unobtainable luxury. They had not been lucky enough to be born into money or into a country with First World standards of medical care. In America, it doesn't really matter much whether or not food or medicine is provided by the particular system we think should be responsible. Some is available to all, and many can aspire to more. But not these people. Not in this country, at this time. For them there seemed little chance that they could escape from a miserable and short existence.

They might have been amazed if they learned that last year, emboldened by the success of my China venture, I had purchased a Lark, a three-wheel electric scooter with a soft, cushioned seat. Together with a new, white Plymouth Voyager van outfitted with a lift to make loading and unloading the scooter easy, the Lark made it simple to run about town on my errands. I wondered if the beggars had ever seen or heard of a Lark scooter or of a van with a lift, or could imagine the freedom they provided.

I wondered if those with eyes that could still see had even seen the monument whose gate they haunted. The Taj was very close. But for someone unable to walk, it was as distant as the full moon that was to shine on it that night. I felt very fortunate and very selfish. I had so much, and they so little. Until I saw those beggars, I had found it easy to tumble into unhappiness by comparing myself with someone more able. But when I saw them, it was hard to see my problems as anything other than small.

Anyone who glanced at the beggars almost immediately looked away. It was hard to meet their eyes or to look at such misery. The sight of them shrieked too stridently against any sane idea of the way the world should be or, worse, could be for anyone if their luck suddenly vanished. It would have been easier for me, like others, to consign them to near invisibility. But, appalled, I could not look away, knowing that if I had been born poor in India the same fate might have been mine.

As T.C. slowly maneuvered my wheelchair through the crowd, they lifted dirty clay bowls for alms when they noticed me watching them. There was no hope, nor even a plea, in the dull brown depths of their eyes. They merely looked resigned.

"Don't give them anything," our driver said. There was annoyance in his voice as, frowning fiercely, he tried to hurry us along. When I looked at him in surprise, he quickly explained: "If one gets rupees, there would be ten more without; if ten get the rupees, there will be a hundred more without; if one hundred, a thousand more; and so on and so on. You will never be able to give enough to make a difference. It's best not to give anything at all."

I waited for a moment when he wasn't watching to give as much as I could to whoever was close enough to get it. Better to be foolish than to be heartless, I thought. Maybe I was not able to help all the beggars in India, but I could at least make things a little easier for a few. But after one dropped rupee precipitated a scratching, slapping scramble, I questioned the wisdom of that decision. I felt very sad, both because of their apparent desperation and because I had inadvertently instigated a fight. The driver turned his head and gave me a knowing look.

Once past the group of cripples, I glanced back. I was startled to see them softly talking with each other and laughing. While making blithe assumptions about the completeness of their misery, I had almost missed the significance of their all having gathered in that one location. Few could have made it on their own. Looking more closely at the group, I noticed several able young boys among them. One was giving water to the old blind man, another spooned mush into the mouth of the one with no arms. In this country, family took care of its own. Almost everyone I met had a brother or sister, son or daughter, or cousin to call on for assistance. Most likely, these helpless-appearing cripples were still included as a necessary part of a family unit and the alms they received contributed like wages to the family pool.

Only several years later did I comprehend how terribly presumptuous it had been to assume that the cripples had no quality of life. Their lives quite likely may have been as terrible as they looked. But perhaps they were not. Without knowing or speaking to them, I had no more right to judge their lives than strangers at home have to judge mine. The only opinion to which I have a right is that beggars exist. I have no say about it. They seem an intrinsic and unremediable part of the population of poor and overcrowded countries. I may not ever understand if they are excluded or merely side-lined in such societies. And I don't know if, struggling to survive, they even care.

These "Children of God," as Gandhi termed them, had been born into the bottom tier of the officially denied, but still operative, caste system. Perhaps, as Hindus, believing that karmic debt from past lives outweighed most present responsibility for one's condition, they had little incentive to analyze their situation or try to change it. To them, fatalistically accepting their lot and functioning the best they could within its framework may have seemed their only option. To achieve a better future they would have to first become victims of the ultimate societal exclusion—death—after which they would get a chance to return and try again in another life.

Soon after leaving the beggars, T.C. slowly pushed me up a nearby hill to view a man who had reached this time of transition. Already he was lying entirely hidden beneath a raggedly ordered pile of leafless branches. With sad faces, well-dressed men calmly performed final rituals for this family member already far beyond any ability to participate. We arrived in time to watch two of the men wedge a few final sticks of wood into his funeral pyre.

After saying he would return in a half-hour and requesting we take no photos, our driver left us on top of the hill. Left alone after T.C. went off to explore, I relaxed in my wheelchair, happy to have a seat for watching the ceremonial activities.

Smiling widely, a slender boy approached. "I take you to where you get good photo," he said.

I shook my head. "No, thanks. I don't want pictures."

His smile got broader. "I can sell you film. What kind you need?"

Finally he accepted the explanation that I wanted to film it only in my mind. "Because of respect," I said.

He nodded. But his eyes were laughing when he said, "I move you closer and explain you what is going on, okay?" He pointed out a third man standing slightly away from the others and bending under the burden of yet more wood. "He will sell his wood to them if they run out," stated my self-appointed guide.

A sudden, pungent smell drifted over on the light breeze as two men shook the contents of a corroded tin of gasoline onto the carefully arranged wood. After lighting a torch and transferring its small bundle of fire to the bundle of wood with the man inside, they jumped back quickly. Suddenly flames shot heavenward. As the fire intensified, the sound of crackling filled the air.

"If it does not burn complete...if there is bone, you know...they will do this again," said my guide. "Perhaps then that man will be able to sell them his wood." Even at a distance, I felt the heat on my face as the red and yellow flames shot higher and higher.

The strangely sweet smell of burning flesh filled my nostrils, smelling much like a baking pig at a summer luau. The boy pointed out a stone gazebo in which almost a dozen men had gathered to sing. With haunting sweetness, their sonorous voices lightly rose and fell with the tones of their plaintively achromatic dirge.

"They will sing for several hours," the boy said.

I nodded. "Where is his wife?" I asked. "Why are there no women there?"

"The wives are away preparing the food," he answered.

The soft breeze cooled the hot tears that rose to my eyes as I listened. As the ancient words rolled gently over me, I thought about how little it matters how hard or easy we have it, or how much we struggle against our fate. Ultimately we all come to this. Such thoughts didn't seem to bother the boy, who, when the time came to leave, waved happily at T.C. as he came to join us.

With strong, willing arms, the boy helped T.C. manage my wheelchair when it threatened to roll out of control down the slippery grass of the hill. Despite the driver's attempts to make him leave, he stayed near, watching for a further chance to insert his help. After loading my wheelchair in the trunk, he stepped back, smiling. As we drove away, he waved so vigorously that his whole body swayed from side to side. Even when we were almost out of sight, his arms were still semaphoring his good-bye.

A short distance away, we asked the driver to drop us off. We were not far from the Taj, but once the driver was out of sight we turned away from it and down a hill. My wheelchair bounced and slipped along a heavily rutted path darkly muddied by excreta. T.C. maneuvered carefully to skirt the vilest puddles as we made our way among a scattering of tiny houses. Adeptly smeared with what appeared to be dung, they looked as if they had a smooth covering of light brown stucco. Previously bright coats of whitewash had been so abraded by monsoon rains that they were now barely more than a memory. As we went by, shadowy faces peered out from the dim interiors of the homes. Several women in dingy saris emerged to get a better view. Nodding and smiling, they pointed us out to naked toddlers whose grimy fingers clung tightly to their skirts.

Everyone watched us, except for one skinny, white-haired man who lay curled up, unmoving, near the side of the path a few feet ahead. As we approached him, I thought he might be sleeping, but when I looked more closely at his milky-yellow skin, I had my doubts. A poor man might sleep in dirt, but I wondered if even the poorest would choose such filth as a place to rest. His chest did not appear to rise or fall. Despite flies crawling on the corners of his eyes, he was utterly still. I did not want to believe he was dead, lying as he was like a forlorn pile of trash. But I could muster no such uncertainty about the condition of the small yellow dog lying near him. With legs stiffly outstretched from its bloat-rounded body, it looked like one of those twist-balloon animals fashioned at children's parties.

Contemplating the cheapness of life in this impoverished village, I felt suddenly somber and chilled—until I became caught up in the bustling community activities that continued on despite the absence of the two lying in the dirt.

Dozens of eyes watched with friendly interest as T.C. struggled to keep my wheelchair from becoming mired in the path's smelly black muck which became wetter and deeper near the bottom of the hill. Barbers, blacksmiths, carpenters, and even men chatting over a mid-afternoon cup of tea all paused. Most smiled and called a greeting. I felt included, at least as much as any American can be in a small, isolated Indian community. If I had walked down that hill, I might have received only a glance, but my rolling down in a wheelchair seemed to be an invitation for more friendly notice. It was an advantage to disability I had not appreciated before.

When we stopped at the bottom of the hill, dark-eyed men, who might have looked fierce if they hadn't been smiling so broadly, ran up to ask if we wanted drinks or food. The offers seemed more a matter of hospitality than sales, but, fearing the sanitation of the refreshments, I turned them down as politely as I could. Soon I was surrounded by a huge mob of children. Nudging, jostling, and teasing each other, the little boys at first stayed a careful distance away and dared each other to touch the wheelchair. Occasionally one of them, made bold by my smile, would break away from the group and dash up to touch a wheel, then dart quickly back to join his laughing friends.

Little girls with hugely beautiful, dark eyes hung back slightly behind the boys. Smiling them, I nodded encouragement. Edging closer, they smiled shyly and gently touched the wheelchair's cushioned arms. Adults watched indulgently. No one scolded the kids for being curious or made harsh gestures to pull them away. Such warnings to children are unfortunately a common practice at home, which makes it difficult for a disabled person to feel included as a normal part of society. Here, however, the wheelchair seemed to be admired by both mothers and children instead of being viewed as an embarrassing or ugly accoutrement.

After checking for my permission and receiving another nod and a smile, the boldest of the little girls cautiously touched, not my wheelchair, but my clothes and red hair. With eyes growing wide as she felt their unfamiliar textures, she whispered excited comments to her friends. They were still pointing at my hair and giggling as we left the village.

Although I saw almost no other wheelchairs in India, my red hair and green eyes seemed to draw as much attention as my manner of ambulation. When notice was taken of the wheelchair, it was usually as a glance of pragmatic assessment or, occasionally, of admiration. Help was offered freely and without any fuss.

To help us get back up the slippery slope, several of the village men grasped the arms of my wheelchair and pulled. "Let me help you, sir," they said to T.C., nodding and smiling encouragingly.

All over India, other slender men were to insert themselves skillfully into positions of assistance: "I'll help you, sir. No trouble"; "May I push you, mem? No problem." A refusal of help was accepted without rancor as long as the helper thought it sincere.

In India, being in a wheelchair was much simpler for me than in America. Help could be accepted simply, because it was offered simply. At home the embarrassed, hurt, irritated, or even angry looks I receive when refusing help with doors or with loading the scooter suggest that I have broken some unspoken contract. Some hidden agenda seems to lurk behind what at first appears to be a simple offer of help. Being helped in India seemed a much more open transaction, and, when I said final good-bye to the villagers, I left their friendly company with more than a tinge of sadness.

Our driver, appalled by our venture into the muddy, impoverished village, told us afterward, "You didn't want to go there." Looking puzzled, he asked, "Why would you want to go there? We have so much nicer things to show you." His voice was wondering. I knew no way to explain how good it had been to be included in village life, if only briefly, and to have the wheelchair so unpretentiously admired.

* * *

On leaving Agra, I was melancholy. During that venture into the neighborhood down the hill, I had enjoyed feeling fully part of life, a celebratory feeling made more pungent by the propinquity of death. But the place and the feeling were soon behind us, and I didn't know if I would ever return to either. My fey mood disappeared quickly, however, on our arrival in Kajuraho when a red mark was gently thumbed onto my forehead. A boy dressed all in white then painted T.C.'s entire face with vibrant primary colors, soon making it into an abstract work of art. We had arrived during the festival of Holi, the rambunctious celebration of the return of spring.

The joy of life was celebrated in Kajuraho in more ways than just by the observance of that brief festival. Near this isolated village was a site where Hindu, Muslim, and Buddhist temples appeared to huddle tightly together. Almost brushing shoulders, they seemed to have drawn close to protect their famous stone carvings. Bands of bas-relief sculptures depicted all aspects of Indian life a thousand years ago and offered an invitation to laugh both with those pictured and at ourselves. Forever preserved in stone, the ancient people almost came to life as they walked, stood, and rode. They seemed all the more real because the Aryan delicacy of their faces still appears in present day faces throughout much of northwestern India. They were shown at work and at play: planting, baking, cooking, and weaving; eating, drinking, dancing, and making love.

I wondered how I would have been depicted if I had been living at that time. Perhaps I'd be the twisted-limbed beggar with a cup or the figure in profile on a bed. If it were a choice, I would rather have been shown in a wheelchair. I was beginning to accept it as an integral companion as I pursued adventure.

In special panels, interposed between more ordinary scenes, erotica was given status equal to more humdrum activities of daily life. Pictured were more combinations of men, women, and even animals than I had ever imagined. Some positions looked as though they would require dislocation of one's limbs to achieve. Some looked like fun. In my favorite, a man was shown pleasing two women while, from a neighboring panel, an elephant, trunk flying to the side, turned his massive head to peer at them in surprise.

Prudish European historians have postulated that the sexual poses, portraying acts described as part of ancient Tantric rites, represented the union of god and goddess when the world was created. When I saw this sculpture, however, no such elaborate reason for its existence seemed necessary. It was enough to discover that its sculptor reached across a thousand years of time to include all of us in his joke.

Sometimes inclusion in all of life's jokes and adventures is delayed because one first has to grow up. This lesson was driven home when I met some of Kahurajo's children. The day I met them grew hotter and hotter as the sun sucked the last remaining moisture out of the village's dusty streets. I, too, felt dried out, ready to crumble and blow away, and at midday retreated to the coolness of my hotel room.

In order to get enough light to read, I opened the curtains and discovered that just on the other side of the sliding glass door was someone's unfenced backyard. Only a few yards away, little girls were loading bricks into a wheelbarrow while a toddler played nearby in the dirt. Spotting me, they chattered excitedly and stopped their work. As I smiled and waved, sounds of their soft giggling came through the glass as they ran closer to get a better view. Even the toddler wobbled closer and gave me a wet, open-mouthed grin.

Intending to meet them halfway, I opened the door, then stopped when I saw the deep ditch between us. It was too wide to cross without jumping, and I motioned that I couldn't get over it. Pointing at themselves and the ditch in imitation of my gestures, they also shook their heads. They couldn't cross either. It appeared we had reached a stalemate.

Two more little girls ran up to join the original four. When I got tired of being stared at and discussed, I waved them away. Soon I discovered that the only way to get them to leave was to close the curtains. I peeked through a small gap where the curtains did not quite meet. After waiting a few minutes, still keeping an eye on the window and the giggling, the girls started to walk away. I had to open the curtains, however, to get light in the room, and as soon as I did they came back. I closed it; they moved away. I opened it; they came back. Another stalemate. It seemed a bribe was in order.

I moved back to search for some granola bars I had packed in my suitcase. I found only two. When I returned to the glass door, the crowd was even bigger. Now there were eight little girls. Holding up both bars, I opened the sliding door and pointed to the biggest girl. Wishing I had more bars, I pantomimed that I would throw them to her and that she should break them up to share. As I again pretended to break the granola bar and distribute it, I pointed a second time to her and to each of the others. Smiling broadly, she nodded understandingly.

When I tossed the bars, first one then the other, my accomplice made two spectacular jumping catches, easily beating out the others who scrambled hard to reach them. Carefully watching her grasping siblings, she made a show of meticulously dividing the bars. After glancing over to make sure I approved, she started to hand them out. She quickly disappeared within a mob of reaching hands. Then, shrieking and running triumphantly away with their pieces, the pack split up. One larger girl had two. The toddler, left crying in the dirt, had none.

As I watched her scrub her teary eyes with grimy fists, I wondered about the wisdom of what I had done and about the meaning of what I had just seen. It seemed like a microcosm of society as a whole. Even if a limited bounty is democratically divided among the many, there always seems to be one who gets double and a weaker one who goes without. I wonder if there is ever a way to fully include the weaker ones in the scramble for a fair share. Sometimes the strong try to even the odds by helping those who are weaker—the young like her, the disabled like me, the old we all one day will be. Yet, over the long term, the scales seem to remain unbalanced unless the weak can develop strengths of their own to act as a counterweight.

As the toddler sat softly sobbing, I looked at her more closely and saw that she was no thinner for her age then her larger, stronger siblings. Like any small child who was to survive, she surely somewhere had a protector—a mother, an aunt, or an older sibling—to make sure she was included and not pushed aside by the others when food was handed out. If she survived through the early years, she would grow enough to get what she needed for herself, both in physical ability and in cleverness. She would no longer be left on the sidelines as her siblings and friends ran after a prize.

I empathized with her physical limitations in obtaining what she wanted, since I also had to look sometimes for assistance from others. Although I could not grow out of a disability as she would grow out of childhood, I could grow in mind and thereby also survive and achieve independence. Yet I still worried about being left behind by the pack—excluded or side-lined from everyday life—until I finally understood what I had learned from the old man holding up traffic on the Delhi–Agra Highway as he walked behind his Brahma bull.

Paying no attention to the rushing life around him, he had seemed content to walk slowly along. If he had seen me at all, I expect he was as little concerned about my assessment of his life as he seemed about the cars racing by. No matter what my opinion, he was going to make his own decisions as to what he needed or desired. And no matter what I thought, he was still going to get where he was going.

The old villager was living proof that there is no need to accept the opinions of others of what you should want or get out of life. When you make your own way, write your own script, it doesn't really matter if you are excluded, included, or somewhere in between. No matter what others perceive as your limitations, as long as you head steadily towards your own goals, you can, like the old man, still get where you're going.

*   *   *

Sometimes, as I discovered when I traveled on to Nepal, even without knowing where you are going or how you are going to get there, you can ultimately wind up in the right place anyhow.

4

Krishna's Boots: Nepal's Magic

Krishna's boots were magnificent. Smoothly covered with yak hair, they were brown and soft and tall. Despite ancient patterns embroidered in red and green on every seam, they appeared masculine and sturdy. They looked like magic boots. With such boots, one might hike to the other side of the Himalayas without succumbing to mortal fatigue. With such boots one might stand against the terrifying Yeti, the "abominable snowman" no one quite dared dismiss as myth. With such boots one might conquer Everest itself during a leisurely afternoon stroll.

Not once, not twice, but three or four times a day, I gazed with yearning at these boots while climbing the steps of the Mount Everest Hotel. The decorative stone rail on which I kept a hand for balance had been designed more for form than for function. About a foot wide, it was curved in a shallow scoop, swooping from four feet high at the bottom to only four inches high at the top. The higher I climbed, the lower I had to bend. Until I straightened at the top and could see his face, my eyes were level with the boots of Krishna, the smiling doorman named for a Hindu god.

Even the simple act of wearing boots like that was now only a dream to me. There was no way my legs could lift their heaviness or my feet could sustain the tiny motions necessary to keep them in place. Every time I saw them I was reminded that now only magic could let me to hike about adventuring. In my former able days I had felt as though I could do it all. But those days seemed now gone.

For almost everyone who encounters bad things in life, "Why me?" is an inevitable question. I was no exception. It discordantly played it over and over in my head like a tune on a scratched record. Then the counter motif would start: "Why not me?" That question was even more difficult to deal with because it required confrontation with a universe no longer as benevolent as it had previously seemed. But as this question rose from pianissimo to forte in the endless refrain of unanswerable questions, it at least demanded acknowledgment. Only then could harmonic resolution be sought or gained.

As my wheelchair bounced over the cobblestones of Kathmandu, I was not intending to start a spiritual quest for an answer to either of these questions. As I soon found out, however, it's difficult not to feel spiritual in Nepal.

At first, Kathmandu—with unpaved, curbless streets, low, blocky buildings, and brightly dressed crowds—seemed much like the many small towns I had seen in India. Much as they did in that country, the people moved with slow purpose amidst scampering goats and placid, hump-backed cows. The only differences about this town high in the Himalayas seemed to be the encircling ring of tall mountains, and the tortuosity of the streets as they struggled over the rugged terrain. That was until I noticed the multitude of street shrines that dotted every intersection.

Housed in each of these miniature temples was a brightly colored statue of a Hindu god in a vibrantly red or golden nest. With flowers placed carefully in front or randomly strewn about, the small shrines brightened the shabby gray-brown of the streets. The light happiness evoked by their presence was blunted, however, by the presence of the beggars grouped nearby. T.C. had to steer my wheelchair carefully to avoid hitting the many people with twisted or missing limbs and various sores and scars. With eyes dulled by resignation, they appeared to have given up any hopes or dreams. As I looked at them I thought, "Why them? Why not me?"

A more unusual sight in a Third World country was the one crippled man who approached us in his equally unusual wheelchair. He was the first disabled person I had seen out in public in this part of the world, who was independently mobile and not begging. The design of his wheelchair was wildly, or perhaps desperately, inventive. What appeared to be a plain, legless wooden chair was sandwiched between two bicycle wheels with grossly overinflated tires bulging around their rims. A complicated web of rusty bicycle chains and sprockets connected the wheels to a lever, which he pulled vigorously back and forth—as if rowing a boat—a few inches above his lap.

As he wheeled rapidly along, the veins in the man's neck and hugely muscled arms bulged with the effort. With practiced hands, he maneuvered quickly around small pyramids of colorful spices laid out on blankets and twined between tables piled high with discarded backpacks and heavy mountaineer's clothing. As he dodged chattering pedestrians, plodding cows, and scampering goats, it became clear that his path was designed purposely to intersect with mine. He seemed all arms and torso as he approached. Giving me a gap-toothed smile, he rolled up and stopped only a few feet away. Only then could I see his legs, or rather what passed for them. Deformed and twisted on his man's body, their size was more appropriate for a ten-year-old boy. Tucked under him, they were barely visible.

Gesturing enthusiastically, he smiled at me and nodded first at my wheelchair, then at his. Through pantomime he indicated that his chair was like mine—only his tires were better. I had to agree. After only a few hours on cobblestones, the slender rigid wheels on my rented wheelchair rode unevenly. When they rattled and shimmied, only T.C.'s determined force could make them steer true. I wasn't strong enough to overcome their warp and move independently. The wide pneumatic tires belonging to my new acquaintance would give a much more comfortable ride and more independence.

I responded with smiles and gestures of my own to indicate that I liked his wheelchair—and that I wished I had his tires. Looking proud, he nodded vigorously. After several more minutes of smiling and gesturing, we parted friends. I would not have met him if we had not shared a similar means of locomotion. Use of a wheelchair was probably, however, our only common bond as we each made our way through life with a disability. For him, with few of my First World advantages, life would be much harder. I could still attain many of my childhood dreams. But if he had any dreams left over from boyhood, I doubted they would come true. Why him? Why not me?

I noticed yet another for whom the chance of dreams coming true had become limited, but his age made it more expected. As we passed in the street, a tremulous old man gave me a wide, toothless smile. His rheumy brown eyes twinkled at the bottom of a pit of many wrinkles. As he nodded a greeting, the thin lanky strings of his long white hair bobbed with the effort. He was so thin that his collarless white shirt and spotless white pants seemed to drape around him like the robe of an ancient prophet. As he hobbled by, clutching his gnarled walking stick with a scaly hand, his pipestem limbs seemed barely able to support him. Finding it equally hard to walk sometimes, and with a head start on many of the complaints of old age, I often felt as ancient as he. Yet, most of us grow old and learn to accept it. There seems no advantage in asking why, because it happens to us all. Yet I still wondered, "Why him?" and "Why me?" But then again, "Why not?"

Later, I saw a man who could have no more dreams. For

him, the end of dreaming had come prematurely. T.C. had hailed a taxi after he grew tired of fighting the shimmy in my wheelchair over the cobblestones. Suddenly, as we approached our hotel, we came upon a half circle of people in the middle of the street. They motioned us around an unmoving young man who lay at their feet, his legs tangled in a bent and broken bicycle. As we drove slowly by, a bright pool of blood was spreading rapidly under his head and beginning to run down the street. As scarlet as the flowers adorning the shrines, his life poured thickly from his slack, open mouth, from a dark gaping hole in his chest and from an obscenely dented skull. Suddenly and brutally dead, he would never reach the destination to which he had too carelessly hurried. As we passed him, I wondered why he was the one hit by a car and why, after several all too near misses, I was the one still alive.

When I thought about those I had seen in Kathmandu—the crippled, the old, and the dead—there didn't seem to be an answer to questions of why me or why them. Pain seems to be a part of life for us all. At some time, in some way, we all get damaged; as babies, our knees get skinned; when we get older, we lose the ones we love; as we age, we lose our youthful health; and, of course, we all die. No matter how good we seem to have it, we can all become suddenly infirm or unexpectedly poor. No matter how remote we wish to think the possibility, pain comes to all of us sooner or later. There seems no way to escape, though most of us wish we could.

Centuries ago a spoiled young prince came to the same conclusion as he traveled around Nepal. He became very disturbed when he, like me, saw the sick, the old, and the dead. However, he also saw a monk, which suggested to him there might be something more to which a man might aspire. He went off to meditate for a very long time, perhaps over many years. Finally he came up with a solution. When he did, a religion was started.

He was called the "Enlightened One"—the "Buddha."

He taught that pain is a part of life only if you live it bound by temporary sensations. As a devout Hindu, he believed that in each life you have a chance to leave that limited self but, if you make mistakes, you pay by returning again and again. However, the hope that inspired his followers came from his teaching that it's possible to stop the cycles from repeating forever; that you can avoid endless reincarnations if you can free your mind from limitations of self. When you can do that you have reached nirvana—a final perfect place.

I had no difficulty with the Buddha's assessment of life, but it was hard to entirely embrace his solution. I could not accept the guilt implied by viewing my present condition as punishment for past mistakes, even if they had been in a hypothetical previous life. Furthermore, I had no desire to spend hours in meditation to free my spirit. Not only did I lack the desire, I lacked the ability. Unable to maintain one position for any length of time, and with sometimes urgent needs of bowel and bladder, I couldn't plan on any prolonged contemplation. I didn't know, or even care, if I had other lives to sort things out. I wanted more instant gratification. So, as I traveled about Nepal, I searched for a spiritual shortcut.

Almost high enough to reach the heavens, this mountainous country filled with shrines seemed a likely place to look for quick enlightenment. Its people, who smiled often at each other and at me, seemed to have found a way to opt out of suffering, even if they couldn't escape life's pain. I tried to learn some of their secrets by absorbing the spiritual atmosphere in their towns and by joining them at some of the places where they worshiped.

First I went to see the goddess who lived in the heart of Kathmandu—the Living Goddess. Now only nine years old, she had been identified as being a deity when she was a mere toddler after being subjected to frightening tests by the priests. In the cleanly swept courtyard outside her home, I waited nearly an hour. A sign announced that the Goddess was available at ten A.M. and two P.M. When asked what she did at other times, the taxi-driver who brought us looked uncertain as he searched for words. "She has goddess duties . . . all day . . . " he said slowly, then looked up and smiled. I wasn't sure whether or not he was serious. Maybe he simply didn't know.

The gathering crowd hummed with activity. Men and women laughed and talked in segregated groups. As I watched a few laughing children run about, I asked if the goddess ever got to come out and play. The driver's eyes widened for a fraction of a second before he answered. His voice gradually gained confidence as he tried to explain, "She plays with selected children, from wealthy families, from royalty . . . you understand? She is not allowed to play running games, just very quiet ones. It is not allowed for her to become hurt."

When I told him I didn't understand how she could reach adulthood without receiving a child's usual cuts and scrapes, he explained further. "She is carried everywhere. Everything is done for her. Her servants protect her." Suddenly, he looked embarrassed. He seemed to struggle for a tactful explanation as he continued. "She will not be goddess anymore if she gets a blemish or if she . . . becomes . . . a woman." He briefly lowered his eyes. I understood. The Living Goddess would be unmasked as an imposter if she became bruised or bled. He seemed relieved when I didn't pursue the topic.

"What happens to her after that happens?" I asked.

He looked sad as he answered, "It is not good for her. Unless she can marry a modern boy, maybe one from a big city like Kathmandu . . . you know?" His eyes crinkled, as if we shared a mutual joke about city boys, then became serious again. "The boys in her village would be afraid to marry her. It maybe would not be safe. She still might have power, could hurt them . . . you understand?" As I nodded that I did, he continued, "Even with modern boys it is hard. No one wants her. She has never cooked or cleaned, done laundry, even dressed herself. Everything is done for her."

He shrugged as he asked, "How could she be a good wife?" Looking satisfied with his own logic, he glanced up at a window behind an ornately latticed balcony. He pointed to the elaborately carved shutters. "That is where she will come," he said. "It is very soon now."

As I stared at the closed shutters, I thought of how my life had some curious parallels with hers. She was as much a prisoner in her ordered role as I was in my disabled body. Each of our societies expected certain behaviors of us, and it was difficult for either of us to break free from the role we had been assigned. Being thought of as a goddess, she would be treated differently than a normal person by those around her, a situation with which I was not unfamiliar. I wondered if she also felt separated from intimacy by the wall of politeness people use to distance themselves from those who are different—almost as if someone's difference had power to hurt them if they got too close.

Although being coddled too much would serve us both ill as we grew older, she would outgrow being a goddess, while I had no chance of outgrowing disability. Perhaps, as she was from a different culture, her expectations were different and my empathy misplaced. There was no way to be sure, since I was only imagining what her life was like. After all, she was a goddess!

Suddenly, in response to some unseen signal, a hush fell over the crowd. Conversations stopped and even the children were still. All eyes focused on the shutters behind the balcony. A minute later, they snapped open. An elaborately dressed young girl with heavily made up eyes peered out: the Living Goddess. Her bright-red mouth showed no pleasure, not even a hint of a smile. She looked very, very beautiful—and very, very bored. She briefly bestowed a sullen glance on the crowd, then quickly withdrew. Her appearance had lasted less than fifteen seconds. As her servants slammed the shutters closed, I knew I would have to seek enlightenment somewhere else.

Perhaps I could find it at the Buddhist temple in a neighboring village. The driver cheerfully put my wheelchair in the trunk of our hired car as he promised to show me everything at that sacred site that those who were not monks were permitted to see. Treacherous with tight switchbacks, the road to the village wound through steep hills. Terraced to fight erosion, they looked as if they had been carved into giant steps. Scanty crops clung desperately to the thin pale dirt on the narrow, slanting ledges. Even the smallest of the gray-brown hills did not look as if they welcomed any life.

Channeled between a steep rock wall and a sheer, precipitous drop, the winding road seemed to occupy almost as narrow a space as allowed for the crops; and its hospitality to life seemed almost as tenuous. When we started to pass two cars, I sharply drew in my breath. It was a blind curve. No opening could be created like the one that had appeared on that flat highway in India. If a car came toward us, there was no way we could manufacture room to squeeze past.

When the pass was completed, the driver pulled quickly back into our own cliff-hugging lane. As my breath came out in an explosive sigh, he glanced back and said, "Don't worry; the eyes of Buddha protect us." It was only then that I noticed two vivid blue eyes watching us from the top of the next hill. Huge and round, their heavy sloping lids were outlined in black. As we curved higher and closer, we were almost always in their view.

When we arrived at the entrance to the sacred compound, I looked around, then gave a deep sigh. Not even Buddha himself could have helped me get to the temple, reachable only by climbing a steep dirt path and several long flights of steps. The driver shrugged when T.C. explained that I couldn't get up there. I was surprised by his reaction. In Nepal, as in India, when I approached a small barrier and needed help, it had been immediately offered. But in the face of this enormous obstacle, the driver seemed indifferent to my disability. Perhaps it was viewed as something to be merely accepted—a matter of fate. Whatever the reason, I was not to approach enlightenment through a visit to that particular Buddhist temple at that time. However, with T.C.'s assistance, I hiked to the base of the trail and sat on a ledge only a few feet below the Buddha's huge blue eyes. Fanning myself with the side of my hand, I watched dozens of visitors walk by.

Several people spun small, bright metal cylinders on short, tube-shaped handles they clutched in their fists. A cheerful man approached me in hopes of making a sale. After demonstrating how to spin the cylinder without it flying off the handle, he handed it carefully to me for closer inspection. Its shape was reminiscent of an ornately decorated, antique baby's rattle. The end of an elaborately inscribed brass handle disappeared into a highly polished copper cylinder ringed with smooth coral cabochons. The free end of the cylinder was loosely capped by a conical top with a slender chain at its point to which was attached a small brass weight.

The salesman gently lifted the top to show me a tightly curled strip of white cloth which almost filled its polished copper housing. He explained that what appeared to be black ink dots soaked in the porous material were the written Sanskrit letters of a prayer. Repeated spins would serve, he promised, to send a constant stream of prayers up to heaven. After I bought one, I practiced the gentle flicks of the wrist he had demonstrated, which started the cylinder spinning. As I watched the more widely spinning nugget on the chain steadily increase the wheel's momentum, I hoped lots of prayers were being spun to heaven and added a few of my own. But, unused to such repetitive exercise, my wrist quickly tired. As my prayer wheel flew off its handle and came apart, I caught the moving cylinder, but the prayer dropped like a whisper onto the dirt.

I hoped for better luck in spinning prayers to heaven when I spotted huge community prayer wheels in the next tiny village. Several feet high, they were lined up in a shallow recess carved into the entire length of a huge block-long temple wall, which dominated the street. Carrying string bags filled with purchases, veiled women gossiped and laughed as they walked by and hit each one in succession.

On the crowded sidewalk along the busy main street, there was no shortage of pedestrians to keep the prayer wheels spinning. Hundreds, perhaps thousands, of prayers were flying up to heaven each minute. To my disappointment, I discovered that, seated in my wheelchair, I could barely touch the bottom of one of the giant wheels. When I stood to spin the wheel, without the forward momentum I could add by walking, the best I could achieve was a hard pat. It only slowed the prayer wheel down.

Despite all the things I had tried, I had found no shortcut to enlightenment. Yet I still was looking for a quick path to escape misfortune. So, I decided to see a fortune-teller. As in many predominantly Hindu countries, such a visit was not difficult to arrange.

While rolling me past a shabby bank of row houses, T.C. spotted a warped wooden shingle which displayed the faded outline of a human palm. In the quiet, almost deserted neighborhood, the narrow buildings looked abandoned. T.C. and I studied the faded sign, hoping for a clue as to where the palmist lived or where he might have gone. Suddenly a high, reedy voice near us asked if we were looking for his father. Although I had heard no one approach, a young, dark-haired boy wearing jeans and sneakers stood close behind, grinning at T.C. and me. With cheerful assurance, he informed us that his father's advice was well respected in Kathmandu, and then, almost dancing with enthusiasm, he led us halfway down the block.

We followed him through a sagging doorway. The bottom of a steep flight of warped wooden stairs was the only thing in the dim, tiny room. Old, peeling paint gave no hint of its original color before it had been covered with multiple layers of grime. After apologizing because his father lived three flights up, the boy stated that the wheelchair would have to remain downstairs and offered his arm to assist me. Swallowing my misgivings, I asked him instead to lead the way.

As I climbed slowly with T.C.'s help, resting every few steps, the boy ran back frequently to check my progress and give encouragement. It was soon apparent why bringing my wheelchair was unfeasible. There was barely room for me to negotiate the tight U-turn at each landing while holding tightly onto T.C.'s arm. Despite the boy's cheerful assurance, repeated every few steps, that there was not much further to go, I began to feel as if I had already been climbing forever.

Finally, we reached a short, dingy hallway. A single light bulb hanging from a chain barely revealed the presence of two narrow doors. Through the one that was open I could see a small spartan room containing only a narrow iron bed and one straight-backed wooden chair. There was not even a single picture to brighten the faded, khaki-colored walls. The boy's smile grew wider as he invited us to go in.

As the first to enter, I looked around, uncertain of where to sit. Knees trembling, I made a quick decision and sat down hard on the thin mattress. Even when I was seated, there was no room for all of us in the room. T.C. hung back to talk to the boy, who waited just outside the door. After assuring me I would not have long to wait, they both went back down the stairs.

A few minutes later, a smiling old man stepped into the room. As his short feathered white hair caught the scant light, it contrasted strikingly with his dark leathery skin. In his undershirt, he was so thin that he had needed to multiply gather his stained khaki pants to fit the excess material under his narrow plastic belt. Seeing the way he was dressed, I wished I were in the chair and not on his bed. But, as he introduced himself, I was reassured by his soft, gentle voice. After moving the chair directly under the light, he sat down and held a ballpoint pen poised above a thin oatmeal colored pad of paper. Looking alertly at my face, he asked my birth date, then bent to write the numbers American fashion, with the month first, then the day and year.

For the next hour his reading of my palm provided him with many more numbers to use in his rapid calculations. As he worked, his assurances, offered in a low and soothing voice, sounded little different than those offered in the horoscope section of almost any American newspaper. His performance was smooth until, in conclusion, along with a discussion of an incorrect zodiac sign, he mentioned the wrong birth date. I corrected him about the month and day, in which he had misplaced the slash then read back using the reversed European fashion of number placement. January 24 (1/24) had become April 12 (12/4) He looked away. For a full minute he said nothing. Avoiding my eyes, he stood and apologized, then told me to come back another time.

For my final attempt at enlightenment, I did something much more direct than merely borrowing someone else's gods, prayers, or rituals. On my last day in Nepal, I flew very near the peak of Mount Everest. Flying seems to be a more spiritual means of transportation than any other. Perhaps it is because, in a plane, the heavens do not appear to be any more distant above than the earth is beneath.

Even this attempt at transcendence initially seemed doomed to failure because of the thick fog that daily obscured the mountains, causing cancellation of flights near the peaks. Finally, a bright pink and gold dawn revealed that the fog had lifted. Freed from their thick, moisture-filled cloak, the huge, jagged mountain peaks surrounding Kathmandu were visible at last. They seemed as unreal as if a gigantic Hollywood set had been slid in overnight behind the familiar mist-shrouded shapes of the closer hills. Suddenly, Kathmandu seemed a very fragile and tiny outpost of civilization when revealed to be at the bottom of such an immense and hostile, spike-rimmed bowl.

The handful of passengers rushed aboard the small jet which was to fly even with the peak at a distance of only twenty miles. I hadn't gotten off the bus in time to board first, as is more usual for disabled passengers, but passengers eager for take-off rushed to help me find my assigned seat. Seconds later, shuddering with effort in the thin air, the small plane took off for the very top of the world.

At cruising altitude the engines developed a high-pitched whine as they bit through air containing fewer molecules than they preferred. From the plane, the Himalayas were distinctive—taller, snowier, sharper, and more extensive than any other mountains in the world. Beyond the nearest peaks, the ranges stretched into a seemingly infinite distance. Squinting against the glare, as we flew down the valley, I was afraid I wouldn't be able to identify Mount Everest among so many other giant peaks. With the help of a map supplied by the Royal Nepalese Airline, I thought I picked out Annapurna and K-2, but I wasn't sure.

Suddenly, all doubts fled when one jagged, black, triangular peak appeared. So tall and sheer that no snow could cling, it stood head and shoulders above its white-covered companions. When confronted with such magnificence, one's soul expands, rushing toward the far horizons of infinity as it attempts to comprehend what is being seen. As my brain relaxed comfortably into the slow beta waves of meditation, I was far beyond disability, not in miles but in mind. It is hard to worry about not walking when you are flying at the top of the world.

During that brief period while the airplane was suspended in the thin air above the Himalayas, I felt whole for the very first time since I had become disabled. The feeling of wholeness came through a sudden surging acceptance of both what my body could do and what it couldn't—the yin and yang of disability, and perhaps of life itself. These opposing aspects are perhaps only more apparent in the disabled because they exist in more sharp contrast than in others. On that brief mountain flight, they joined for just long enough to show that they are both part of one single entity. In the joining of the two parts—an acceptance of life on life's terms—the wholeness is complete, and one can find, if not nirvana, a sense of inner peace.

There was no way to understand why I was so privileged as to be on that flight—no way to force or borrow an answer to the unanswerable. Questions of "Why me? Why am I the one that's so lucky?" had no more metaphysical explanation than their opposite counterparts. In the face of the overwhelming beauty of the mountains, any remaining concerns about answers to those and similar questions disappeared into the infinite distance. No answer had been needed to achieve wholeness.

I would have liked to have physically challenged Everest's peak, although, even in able days, that had probably never been a realistic option. Now, as a wheelchair traveler, I conquered it anyway. I didn't do it with my body. I did it in my heart and with my mind. That is how I finally learned the secret of magic—the magic of Krishna's boots.

*   *   *

Perhaps the magic of Krishna's boots was what was needed to solve the problems of Russia. When I went there in 1991, a year after leaving India and Nepal, it seemed that only something on such a giant scale would do.

5

Gaining a Sense of Proportion in Russia

Within minutes of my entering the Republic of Russia, a loud, vociferous dispute erupted between the two Russians who met us. They were vigorously arguing about me or, more precisely, my wheelchair. Lara, the smiling Intourist guide, had not seemed perturbed when I arrived in it, but that was not the case with the thin man in a navy-blue jumpsuit who waited for us just outside the terminal.

After glancing at me, he shook his head vigorously and spat out a few abrupt Russian words. Even his brush-cut, gray hair seemed to stand straight up with indignation as, with mouth mulishly set, he continued to move his head in stubborn negation.

Lara appeared visibly upset as she argued urgently with this man who had been assigned to drive us to our Moscow hotel. Her chin jutted aggressively forward and her voice became increasingly strident. As her buxom chest swelled with anger, her well-tailored leather jacket gaped awkwardly above its loosely tied belt. Her fashionably short, blond-streaked hair bounced in synchrony with the increasingly vehement chopping motions of her hands. She looked very intimidating, but the driver appeared unmoved by the force of her words.

Avoiding her eyes and mine, he stared into space. His arms, thin as pipe cleaners, were crossed tightly across his scrawny chest. Occasionally, he uttered a string of harsh-sounding words and briefly freed his right arm to gesture roughly at the bright-red wheelchair I had bought for this trip. It wasn't hard to understand what was going on. My wheelchair would not fit in the trunk of his car with the luggage, and the driver was insisting it be left behind. Oblivious to its importance, he saw it merely as excess baggage. Although there was ample room for it in the trunk if T.C. and I shared the back seat with some of our bags, it appeared that he could not accept any deviation from the usual way of distributing passengers and luggage.

Finally the argument escalated to a point where he shrugged his shoulders and stalked to the driver's seat, ready to leave without us. Lara followed closely on his heels, promising that in return for being allowed to load our own luggage and share the back seat with some of our bags, we could be counted on for a sizable tip. Although she now displayed a placating smile, her tone of voice was still one of command. She did not seem like one who was used to being disobeyed.

"Don't worry," she said to me. "I have a college degree and speak several languages. He is only a driver."

There was a brief silence as we watched him reconsider. Glazed with concentration, his eyes shifted from side to side. The gravity of his reaction, even the argument itself, seemed way out of proportion to the situation. Finally, he gave an abrupt nod and motioned for T.C. to bring the wheelchair to the rear of the car.

That such a large argument could erupt over such a small problem was the first indication that, in order to make sense of Russia, one had to understand how disproportionate things were in that republic. Most of us instinctively understand the importance of proportion—the relationships of things—in our everyday life. We try to keep things in proportion, we suffer when we blow them out of proportion, and we worry when someone gets a disproportionate share of something. When things become difficult, keeping a sense of proportion helps us to avoid feeling overwhelmed and lets us smooth out our lives. The Russian Republic seemed overwhelming at first, because the proportions of things in it were more exaggerated than those with which I was familiar.

After being dormant for decades, in early 1991, the giant Russian bear, prodded by *glasnost* and *perestroika*, was starting to stir in its sleep. As the resulting tremors spread across the land, the Russian people still had hopes of controlling the wakening behemoth.

"Yeltsin is our man. We don't need Gorbie," said one smiling woman I met in a Moscow store.

"We like Yeltsin. He's cute," giggled two teen-age girls who were our guides in Siberia.

"We are Russian; we don't need the other republics. We have most of the land and most of the riches. Why should we care about them?" said a waiter at a Leningrad hotel.

Things were not, however, to be as simple as they hoped. Nothing about Russia was simple. No matter who was in power, the bear was not going to suddenly become tame or quietly go back into hibernation. The USSR was very soon going to come apart.

As a consequence of their leaders' maneuverings, Russians faced a future of almost overwhelming uncertainty. The ultimate effects of economic and social changes now just beginning lurked like giant sharks underneath the uneasy surface of Soviet society, ready to pull the unwary under. I could only glimpse the ripples on the surface and guess at what huge forces might lurk beneath.

On the drive from the airport, the large scale of Moscow's streets and buildings hinted at the enormous controlling force hidden beneath the superficially placid surface. The city seemed designed more to impress visitors with the authority it housed than to provide a comfortable place to live and work.

The broad, rain-shiny streets seemed even wider because of the paucity of traffic. The buildings we passed—high-fenced universities, grimly utilitarian offices, and rococo churches with red stars topping their steeples instead of crosses—towered over the few heavy-coated pedestrians who walked briskly along the wide sidewalks. When we arrived at the Intourist Hotel, a spartan concrete block with narrow windows arranged in grimly precise rows, it seemed neither more welcoming nor more potentially comfortable than a fortress.

It was hard to imagine how the Russians, dwarfed and diminished by their cities as well as by a large array of new social problems, could hope to attain identity and power without a struggle. It was even hard to be sure that simple human needs would be met in such uncertain times. Caught by forces out of my control, I was also struggling for personal identity and power while trying to meet changed needs, and felt an unexpected identification with the Russians as they worked hard to have their voices heard.

Inside the Intourist Hotel, the diminution of people in Moscow by the grand scale of their architecture took on a more immediate and personal meaning. It was not unusual that the counter of the check-in desk was several inches above my head. Counters everywhere—those used for airline check-in, bank transactions, and customer assistance, for example—are usually above eye level for anyone seated in a wheelchair. What was unusual was that the guest rooms, with very high ceilings, elevated beds, and oversized chairs, all seemed to have been designed for a race of giants.

Even the dining hall was so large that voices echoed hollowly through its vast, empty spaces, as if reverberating in a huge, empty cave. Broad tables in widely separated booths seemed more appropriate for playing ping-pong than for dining.

When I sat on the built-in bench of the booth we were assigned, its twin across the table seemed as distant as an opposite bleacher in a stadium. My feet, like those of a four-year-old child, barely touched the floor, and my arms were barely long enough to reach my coffee cup.

"Real coffee is hard to get. I told them to make it special for you because you are VIP," said Lara, tossing her head proudly, after we were served. I felt more like Alice in Wonderland after she drank the potion that made her small.

The feeling that somehow I had shrunk continued the next day at Red Square. Perhaps to a person striding purposefully across its gray-brick expanse, it would seem just an average-size town-meeting place. But to me, as I bounced over its uneven surface in my not-well-enough-cushioned wheelchair, it seemed enormous. Without a way to relate its proportions to anything with which I was familiar, it was difficult to judge how big it really was until I thought of the May Day celebrations beamed each year to American television. On that Russian equivalent of the Fourth of July, huge crowds seemed easily contained in the square as they watched troops pass in review under the steely eyes of the old men they served.

Small plaques embedded in the Kremlin wall commemorated many of those grimly powerful men, past chairmen of the Communist Party. To my surprise, I discovered the little square plates were also their grave markers. For all of them except Lenin—for Stalin, Khrushchev, Brezhnev, Malenkov—the huge walls of their citadel of power had become the walls of their tomb. The smallness of the embedded markers seemed in no way proportionate to the enormity of power wielded by these men whose slightest whim had brought death to so many.

In sharp contrast to these meager markers were the elaborate and fanciful memorials located in a park-like cemetery nearby. These memorials paid tribute to better-loved figures, such as poets, authors, and even Khrushchev's wife. Above their graves, laughing limestone children huddled on gray, granite obelisks and contemplative stone women posed seductively on long, slender sarcophagi. Frozen forever in bas-relief, serious-faced bronze men stood watch over all. Multitudes of such monuments were gracefully strewn about, sharing the neatly mowed lawns with scattered clots of brightly colored wildflowers.

This peaceful, pretty cemetery was a favorite place for slender, blond brides to come frothily gowned in ruffled white organza to pose with their more sedate young grooms for wedding photos. The love they expressed as they stood smiling with their friends on the grass seemed in keeping with that displayed for long-ago brides and grooms by those who had placed bouquets at the bases of the elaborate sculptures.

The sharp contrast between the markers of gravesites here and on the Kremlin wall may have been a reminder of how differently party leaders were to be remembered. The disproportionately small size of the markers for the once enormously powerful people buried in the huge Kremlin walls perhaps carried the message that it was the power of the party, not the party chairman, that was to be remembered.

Despite the enormity of secular power they symbolized, the Kremlin walls were overshadowed by nearby St. Basil's Cathedral. Towering high above Red Square, the cathedral dominated its surroundings. Silhouetted against gray skies, low and heavy with impending rain, the cathedral was the only spot of color in the colorfully named square.

St. Basil's garishly painted onion domes looked like large, fancifully wrapped Hershey's kisses as they squatted placidly on red-brick towers of varying heights. As they circled a fat steeple topped by a small golden dome, the vertical stripes of blue and white on one competed for attention with the wavy stripes of yellow and green on another. Looking like a pregnant barber pole, one very large red-and-white striped dome clashed incongruously with all the rest.

Although the architecture of St. Basil's was somewhat unusual to a Western eye, the tall, elongated crosses added to the pointed tops of the domes made their proportions somewhat more pleasing. Slender and golden, each cruciate spire stretched heavenward like the finger of Michelangelo's Adam eagerly reaching to touch the finger of God.

The strength of St. Basil's visual impact, however, was far out of proportion to the current power of the orthodoxy which it had served, a power now emasculated by mandated atheism. Although not apparent from a distance, the sacredness of the vividly colored cathedral had been profaned by the partitioning of its interior to house government offices.

Worried that rivals might commission other cathedrals to outdo this masterpiece, Ivan the Terrible had the architect's eyes put out when it was completed. The talented artist had been left with only his inner vision. Even now, the cathedral seemed barely more than a fantasy, more illusion than real.

Only a few hundred yards away Lenin's tomb also seemed surreal. After a soft guttural command from an officer, four Red Army soldiers smoothly lifted me in my wheelchair, then carried me effortlessly into the heavily guarded mausoleum. As they somberly marched in perfect step up the short flight of stairs, then silently around the bier where Lenin's body lay in state, it seemed almost as if I were part of his long-ago funeral procession.

In the softly illuminated room, only a very slight shimmer around the silent, recumbent figure hinted at the bullet-proof glass around this man who had started a revolution.

"He is in a chamber from which the air has been taken out," Lara whispered, then quietly pointed out temperature controls that kept him even cooler than the rest of the shadowy room. No one was willing to take a chance on the reliability of the decades-old embalming fluid that presumably still filled his long-dead arteries and veins.

The lights reflected off Lenin's round, balding scalp much as they would off a piece of waxed fruit. His surprisingly red goatee, with each individual hair distinct, looked freshly brushed as it jutted forward in the distinctive shape made famous on posters and banners. Looking almost too perfect to be real, the body resembled a wax figure waiting to be shipped to Madame Tussaud's famous museum in London. The peacefully closed eyelids and carefully shut lips gave no hint of the fire in eyes and tongue that had sparked a conflagration powerful enough to consume an empire and lead to nearly a century of bloodshed and fear.

Such a small man. Such a gigantic legacy. What a disproportionate influence he had had on the world. Thinking of it, I felt cold, but I could not stop staring at him even while the platoon carried me smoothly down the few steps to the exit. Although it was drizzling outside, the cloud-filtered sunshine offered welcome relief from the pervasive chill of the tomb.

As Lara, T.C., and I started to cross the square, a scowling soldier brought us to a sudden halt. Eyes narrowed with hostility, he indicated with bitten off words and abrupt hand motions that we were not to go any further in the now-deserted square. To my surprise, Lara planted her feet and appeared ready for a dispute. Gesturing vigorously towards my wheelchair, she began to speak in firm but coaxing tones.

The brightly polished red star on the soldier's cap glistened with authority as he occasionally glanced from her to me. His scowl slowly faded and his tense muscles gradually became more relaxed. Finally he shrugged and stepped back, motioning his permission to cross the square. As she hurriedly led the way to our hired car, Lara directed T.C. to push me quickly. She breathlessly explained that the square was being cleared for security reasons. A VIP and his entourage were expected in the immediate future. She joked that I was a VIP, too.

Presumed helpless and harmless in a wheelchair, my presence had brought us special privileges—as long as we were quick about it. Moving rapidly over the rough, gray bricks, I held on very tightly to my armrests to keep from being bounced from my seat. In view of the rough ride, I was not sure that receiving what an able-bodied person would view as a special privilege was anywhere near as desirable as being treated the same as everyone else.

Even special privileges would have brought little advantage when we went a few blocks to shop at GUM. The elaborate building looked more like a gigantic American mall than the simple department store it was proclaimed to be. Tall, sweeping staircases drew the eye up to the intricately filigreed wrought-iron railing, which edged the huge stairwell on the second floor. Then the eye was drawn up again, then yet again, and finally to an elaborately paned glass roof more than four stories above.

The building's huge size and classical formality seemed way out of proportion to the extent of its present use. At first glance, there seemed to be nothing unusual about the elaborate displays of merchandise sparkling behind huge, glass display windows. There were medicines in crisp, modern packaging, Parisian perfumes in delicately sculpted flacons, and new clothes, curiously similar in style to those of Depression-era America. On closer inspection, however, it was apparent that each careful arrangement of displayed items was designed to disguise the fact that there were very few of them. Inside the shops, sparkling metal shelves displayed more empty space than merchandise. In this largest and most famous shopping place in Moscow, there was very little to buy.

Strolling window shoppers seemed dwarfed by the stores as they slowly wandered through the wide aisles. None seemed to be going into the shops. In front of one row of stores stood a long line of Muscovites. Appearing overheated in long, heavy, military-style coats and bulky astrakhan hats, a few bored-looking men shifted heavily from foot to foot. They were greatly outnumbered by dozens of women in heavy, dark coats and long woolen stockings. An unusual number of the patiently waiting women had dark red hair peeking out from under large furry tams. Where the light was strong, it was apparent that the odd blood-clot tint of their hair was caused by poorly applied henna. Lara explained that many women used the vegetable dye to keep their hair from falling out during the long, sunless winter.

When I suggested it might be more helpful to take vitamins, Lara quickly insisted that the farms in Russia's more temperate areas produced an abundant supply of fresh vegetables all through the winter—so ample they didn't even need to import any from the Ukraine. With a defensiveness quite out of proportion to what I had considered a simple medical observation, she asserted again that the hair loss was a mysterious problem needing only henna for a cure. I wasn't sure if she was defending her republic's ability to produce food or covering her own lack of knowledge. Her mouth showed a hint of stubborn tightness, and I didn't try to argue.

As they stood desultorily chatting, some of the waiting shoppers pulled out packets of food. It appeared that they were settling in for a long wait in the non-moving line. When I asked Lara what was being sold, she walked up the line and asked each person in turn. Receiving a number of negative head shakes, she moved up the line, then disappeared with it around a corner. Soon she returned, puffing with exertion.

"It's men's winter boots," she said.

"Are they having a spring sale?" I asked, before I remembered what I had heard about Russia's shortages of consumer goods. I compounded the error by naively adding, "How can women get a fit in boots made for men?"

In a tone which implied I should understand the obvious, Lara said, "But the boots are here now. They buy what they can get. Each is allowed one pair. Fit doesn't matter. Maybe they can give to a relative, or stuff with paper." She paused, looking speculative. "Maybe they trade for something else... maybe sell, make a little profit. They will all stay until the store runs out of boots."

In Russia, despite its Communist ethic, no one seemed to contend the propriety of having the best goods reserved for only a favored few—for those in power and those with money. Despite the scarcity of merchandise in the most frequented stores, a disproportionate glut of goodies filled special stores reserved for tourists. There, only foreigners with dollars, pounds, francs, marks or yen—hard currencies much sounder than the ruble—could legally make a purchase.

For visitors, an abundance of expensive foreign items as well as elaborate local crafts were available. Tall stacks of high-priced foreign liquors and the American cigarettes much coveted by the locals immediately drew the eye. Mink and sable hats competed for space on long, deep shelves with ornate balalaikas and samovars. Glossy-black, lacquer boxes, painted with colorfully clothed princes and magicians, were crowded together with matrioshka nesting dolls, "little mothers" containing up to eighteen miniatures of themselves one inside the other. Most of the souvenirs were reminders of the old Czarist Russia of which I had hoped to catch a glimpse. But, except for symbolic trinkets in tourist shops and a few well-guarded exhibits in museums, it seemed what I sought was now only a memory.

Yet hope dies hard. By making a side trip deep into the forests of Siberia, I made an attempt to find out if any of that old culture had been preserved. The small central Asian town of Irkutsk, isolated from Moscow by the giant Ural Mountains, seemed a good place to start. Even in the era of jet travel, it was hard to reach and seemed far enough away from the capital that it might be only loosely gripped by the grasping tentacles of centralized power far away in Moscow.

* * *

Barely skimming the tall, hostile mountain barrier, our Aeroflot plane struggled though air thinned by high altitude as we headed deep into central Asia. After wiping away an icy condensate on the window with loose stuffing pulled from a slit in the aircraft's wall, I pressed my face against the cold glass. Row upon row of white, spiky mountain peaks looked like a nest of sleeping dragons—not the friendly playmates of children's stories, but treasure-guarding monsters of an older mythology.

Bleak and dangerous, the sight of those endless ranges of treeless peaks rolling into infinity was even more chilling than the cold of the icy window. After the plane finally crossed the Urals, however, the dark, dense, inhospitable forests covering low, rugged hills were hardly more reassuring.

Perhaps that was why the plane's cabin suddenly seemed a cozy home and the strangers aboard like comforting friends. There was a feeling of comraderie, an instinctive acknowledgment of our need for each other in the harsh environment we were soon to enter. It did not, therefore, seem unusual when, shortly after our plane landed in Irkutsk, T.C. received a grin and a salute from a disembarking Soviet Marine. Only later did I ask why the man with the bluff, chapped face and twinkling blue eyes had lightly touched his fingertips to his black-banded cap as he passed our seats. As calmly as if it were a usual event, T.C. said that the officer had helped him cram my wheelchair into the baggage hold after the ground crew in Moscow had refused.

From the air, Irkutsk had appeared only to be a small dot in the middle of the wilderness, and the town didn't seem much bigger when we were driven through its streets. Tiny homes seemed to huddle together as much for protection from the surrounding wildness as from the extremes of winter cold. The intimidating terrain surrounding the small town seemed to bear witness to man's insignificance, but the citizens of Irkutsk, descendants of Stalin's prisoners, found their own ways to bring their world into more reassuring proportion.

Many people had made their warm-looking log cabin homes even more welcoming by adding ornately carved wooden balconies and shutters. Warm, golden light pouring from the windows and colorful splashes of flowers spilling from prettily carved window-boxes added further coziness to the scene. The town's cheerful atmosphere did not, however, result only from the presence of these inviting homes. Many of the people walking slowly down the main street with overflowing shopping baskets paused to smile broadly and give friendly nods and waves.

Our two eighteen-year-old Intourist guides merrily informed us that the severe winter cold wasn't so bad if you were born there. Although their fathers had been sent there as prisoners, the young women assured us that they were quite happy to live far from Moscow and be left to their own style of life. As they explained the features of their town, they giggled frequently, making their pony-tails quiver and their pom-pom hair-bows bounce. Life in Siberia was pretty good, they told us, except maybe . . . giggle . . . for the lack of enough boys their own age.

In a surprisingly short time, we reached the outskirts of town, where clusters of tidy homes and stores were quickly replaced by unbroken forest. As we made our way to Lake Baikal, the area's biggest attraction, tall rows of slender white birches extended as far as the eye could see. As it twisted among the ghostly trees, the road seemed as slender as a thread, an insignificant afterthought. Ultimately the birches were replaced by trees of a less familiar type, packed close to the road like a tight, impenetrable fence. The road now seemed proportionately wider as it forced its way through the narrower space. Only a tiny glint of blue seen through the dense branches gave any hint that we were approaching the deepest fresh-water lake in the world.

That tiny glint was nearly all I got to see of it. The vantage point chosen by Intourist was from within a restaurant perched high on a hill, and no one would help T.C. carry me up the several steep flights of stairs which were the only means of entry. The driver, who had willingly loaded my wheelchair into the trunk of his car, seemed surprised there was a problem. After the guides giggled and explained, he shrugged his shoulders, then, shaking his head, grimaced and grabbed the small of his back. He gave me a quick glance of sympathy and hurried off to his lunch.

In the harsh Siberian environment, a visitor's weakness was not to be the cause of another going without food. A few minutes later, the girl guides and T.C. left me in the parking area while they went for their own lunch, with instructions to bring back a doggy-bag.

I settled back to enjoy the weak sun and the crisp, clean air. With just the right tilt of the head, an almost incandescently blue triangle of lake could be seen through a gap in the surrounding trees. The silence was as total as that in a cathedral. I never found out why in Russian folk songs it was lauded as the "Holy Lake Baikal," but it was understandable why some might feel they were in a sacred place.

*   *   *

It was not until we returned to Moscow that I understood why my use of a wheelchair seemed so misunderstood by the

Intourist drivers. On one sunny Sunday, Lara proudly explained that Russians were once again allowed to celebrate religious holidays. Handing T.C. and me small, freshly cut elm branches, she invited us to her own church to share the holiday by placing the greenery on the altar. Soon we pulled up at a Victorian-style church that, like most places of worship in Moscow, had a red star atop its steeple instead of a cross.

Very near the church entrance, eight or ten shabbily dressed men sat on the cold concrete sidewalk. Their clothes, threadbare and dirty, did not look much warmer than the rags worn by the beggars in much warmer India. Many of the grizzled men were missing limbs; some had grotesque scars. Some, with deeply wrinkled faces, displayed rusty medals on frayed and faded ribbons. Perhaps, like the men honored by the beautiful, sculpted statues scattered around the city, they had fought bloody battles with advancing German troops during the Second World War.

With pleading eyes, they wordlessly held up grimy, tattered caps, trying to draw the attention of the church-goers. A few people dropped kopecs, each worth less than a third of a cent, into the upturned caps before quickly turning away.

As T.C. pushed me by, I was startled when I felt the light brush of someone's hand against mine, followed by the cold hardness of a small coin in my palm. Until I gathered my wits and clasped my hands firmly together on my lap, several more kopeks were pressed into my hands. I felt strangely guilty, as if I had stolen them from those to whom they rightfully belonged. Yet, I also felt like something had been stolen from me. By being the object of charity, I had lost some of my dignity.

The coins burned my palm. When we left, I hurried to drop them into the caps of those still waiting outside. But the act did not bring as much relief as expected. By assuaging my conscience, I had diminished them, too, unless the last of their dignity had already been stolen away.

I saw a few other disabled people in Russia and had been variously told that it was because people were ashamed of them, that all the stairs in the cities made it impossible for them to get about outside their homes, or that they were all happily living in group homes in the country. Since there were so very few other disabled people around, none of them Russian, it was hard to know what to believe.

When I saw the beggars at the church, the full meaning of the word "handicapped" suddenly hit me with the full malevolence of its medieval definition: "cap in hand." I had occasionally used the word to describe myself, following then common custom, but now the term was far less appealing. Yet it was hard to think of a more acceptable term.

One woman in my hometown, whom I had phoned with a question about the accessibility of a store, had bridled when I'd told her I was handicapped.

"Surely there must be some other term you could use," she had snapped. "Isn't the way we say it now 'something-impaired'? Which kind are you?"

Stuck for an answer that wouldn't offend my sense of privacy, I had soon ended the conversation. Mobility-impaired? Balance-impaired? Energy-impaired? Sensation-impaired? Bladder-impaired? Concentration-impaired? What?

"Crip," said T.C., when I asked him that night at dinner after we had seen the beggars at the Moscow church.

"That's a terrible term," I said, thinking, *Crip, short for crippled—short for useless.* I hated the sound of it.

T.C. tried to explain that that's what the "in crowd" called each other—the wheelchair racers and basketball players, the movers and shakers in the disabled world. But I would have none of it. To me the term was as inflammatory as a red flag waved in front of a bull, and not to be tolerated.

But what should I call myself? What shorthand description would be useful to indicate needs for access or for rest? "Disabled" had too close a connection with a junked vehicle by the side of the road, and "MS-er" seemed to imply that a disease was my whole identity. "Handi-capable" was just too cute, "differently abled"—too bulky and not really true. Except for difficulty in walking, I had the same abilities as everyone else.

I now solve the problem by using many terms. In speaking to others, I sometimes use "handicapped"; it is probably the most universally understood until we can retire the term. When writing, I use "disabled" because it takes less time to type. Privately, I now call myself "crip." The terms do not define me. I define them. My use of such adjectives to describe myself acknowledges that physical impairment is an integral part of me, as inextricable as my skeleton and as personal as my name. My use of any of these terms is a description of an identity I choose to reveal, and, thus, a statement of personal power.

*   *   *

In Leningrad (now St. Petersburg), I found another who derived a base for power from physical impairment—not his own, but that of others. Like many other Russians, Igor was seeking a secure role in his changing society. By operating the only lift-van in Leningrad, perhaps in all of Russia, he was carving out his own niche in tourism. Because of his monopoly on the service, most wheelchair travelers to Leningrad would travel in his special vehicle outfitted with a lift which picked them up from the street in their wheelchair and brought them back down at their destination.

Like many young Russian men, Igor was so thin he looked almost emaciated. His sallow skin and brittle mouse-brown hair made no secret of chronically poor nutrition. When he smiled, which was often, he displayed decayed and broken yellow teeth. Shortly after we met him, Igor proudly demonstrated how he worked his lift. Although his beaming explanations, translated by the guide, were enthusiastic, his scanty muscles seemed barely adequate for the task. Dripping with sweat and grunting with effort, he slowly turned a hand-crank to raise the platform holding me in my wheelchair to door level in his van. His smile was so brilliant when I was finally inside, I didn't have the heart to tell him that I would have preferred to climb in myself and have the wheelchair brought in separately.

Like Lara in Moscow, who with a few friends planned to compete with Intourist when the government gave up its monopoly on tourism, Igor seemed to have the necessary entrepreneurial spirit for success in a society new to capitalistic thought. Happiness for both of them, however, would depend on their satisfaction with the relative proportions of their roles: Lara as a small fish in a big pond; Igor as a big fish in a small one.

As Igor sped through the wandering streets of Leningrad, many of them tortuously fashioned where canals once had been, it was again noticeable how much Russians were diminished by their cities. Leningrad was another city drawn on a gigantic scale, although its buildings seemed more welcoming than the gray utilitarian ones in Moscow. This was due in part to ornately curlicued window frames, which lessened the austerity of the pale-yellow, neo-Classical buildings. It was also due to an abundance of green parks and to the remaining canals that flowed peacefully amongst them. For me, however, the main attraction was that the huge rooms and galleries in the Hermitage and Winter Palace were accessible by elevator, unlike those in other city museums.

We spent our last day in Russia viewing part of the incredible legacy of art left by the czars. At the entrance to the Winter

Palace, one of their former homes, was a large sign requesting visitors to remove their shoes. Nearby, shelves were partially filled with footwear of all kinds and high stacks of paper slippers. I carefully watched the eyes of the guard for disapproval when he noticed the dirt-caked wheels of my wheelchair. Yet, for reasons I did not quite understand, he impatiently waved me toward the elevator—dirty shoes, dirty wheels, and all.

A seeming endless array of gilt and crystal was presented in room after room in the Winter Palace, each room more fantastic than the last. Sitting rooms, bedrooms, a music room, a dining room, and, of course, a throne room looked as if the royal family had just stepped out for a moment. The rich lifestyle of the czars, as indicated by the expensive art collection and the plethora of household objects made of gold and heavy silver, had been way out of proportion to the poor conditions borne by their subjects. Something that far out of proportion had made many resentful and uneasy, just waiting for a Lenin to emerge.

Ultimately we crossed into the Hermitage, where thousands of Great Masters are housed. The huge size of the museum precludes anyone from seeing more than a small sample, but once I had seen one of da Vinci's madonnas I needed to see no more.

In the dim, cloudy daylight creeping through the small windows, the madonna's almost translucent skin seemed to glow as she gazed at the tiny infant in her lap. Her perfect lips curved in a gentle smile of sheer delight as the chubby baby, eyes wide with wonder, reached to grasp a curly-petaled, purple flower she held.

Time, space, even physical existence became meaningless as I stared at them. I forgot I was in the twentieth century. I forgot I was in a museum. I even forgot I was in a wheelchair. All I was aware of was the woman and her baby. The beauty of the painting was overwhelming. I wondered how this mother and child, who had had so giant an impact on the world, could be portrayed so realistically in such a tiny space.

As I struggled to make sense out of what I saw in Russia, I learned that both the size and importance of things—countries, buildings, people, and even our own problems—seem to be determined by the context in which you view them and by how you interpret what you see. These two things almost always determine the choices we make while attempting to put things into proportion, something we need to do to create a good life for ourselves. And we do have choices, no matter what our situation. We can even choose what factors we want to consider in making our choices—ignoring some, expanding others. It doesn't always come easy, but if we choose well, we may be well on our way to understanding the secret of happiness. At the least, we can make enough sense out of things to be content; and that is a fine way to live a life.

* * *

A year after leaving Russia, however, I was involved in the battle of my life as I tried to keep things in proportion when they were suddenly blown far out of it.

6

And Now for Something Completely Different: The One Journey I Would Rather Not Have Taken

"It's adenocarcinoma." The softly spoken sentence shot like a lightning bolt through the clouds of anaesthesia that mistily shrouded my mind. I had been so sure the pathologist's report would be negative. Adenocarcinoma. Cancer. Breast cancer. The big "C"! Fear spun around me as if I were a pilot in a terminal stall and spiraling straight down into a jungle. There was little time to gather my wits about me as I was suddenly plunged into a journey I would rather not have taken.

When one crash-lands in an unknown land, shock replaces the initial heart-thumping fear while the mind denies the situation. But as awareness seeps in, you must take stock of where you are, then quickly go about the business of survival.

You know you need to move fast before you are overcome by the elements or the hostile terrain. But at first, you slog through the unfamiliar territory slowly, fearing that your next step might land you in deadly quicksand.

Your survival depends on gathering as much information as you can. In a strange jungle fraught with danger there is no reliable guidebook, no safe passage, and no guarantee of getting out alive. Hoping against hope, you desperately look for an experienced native to teach you how to make weapons, warn you of pitfalls, and, most of all, show you a way out.

This expedition was far more dangerous than any I had made previously. It was also different than any of the others because I neither planned it nor found it interesting. Most of all, it was different because I couldn't wait to get away from it.

One way of gaining control in a hostile jungle is to clear it. In similar fashion, to gain control of cancer, it must be slashed, burned, and poisoned. Unfortunately, the reluctant traveler has to undergo the surgery, radiation, and chemotherapy along with the wayward growth. For me the slashing came one warm day in June, when a mastectomy relieved me of my tumor-containing left breast. After it was decided that burning was not necessary, I dutifully reported for poisoning.

For the next six months, I was treated with chemotherapeutic drugs—the classic three known casually as CMF: cytoxin, methotrexate, and 5-fluorouracil. Despite horror tales willingly supplied by strangers, chemotherapy was really not all that bad. Like many women, I had no hair loss or nausea, just a feeling of general fatigue. After living with MS for ten years, I was already used to that, so it was not as impressive as it might have been. The only real problem with the poisoning was that I had to go to the office of a cancer specialist to get it.

For those who have crash-landed in the jungle, entering the office of an oncologist gives as much a feeling of helplessness as when a tropical snake grabs hold of your body, then swallows you. After a hesitant approach, you are suddenly surrounded by coils of bureaucracy, which keep you immobilized in a waiting area for a fear-filled infinity of time. After your predecessor is finally digested, a gaping maw opens briefly: "You can come in now," a voice tells you as an insistent peristalsis urges you along the wake of a rapidly moving nurse.

Sharp little teeth chew briefly on the soft inner flesh of your out-flung arm as various components of your blood are sampled. After a few brief squeezes to check your blood pressure and pulse, you continue moving along until you come to a halt in a small inner chamber. There, after a few investigatory prods of your quivering corpus, you hear a doctor's voice say, "See you in two weeks," and the regurgitation process begins.

At first, you are only partially expelled so venom, missed on entry, can be injected. Finally, filled with toxins, you're vomited out, promising to come back for more.

One way in which fighting cancer differs from fighting one's way through a hostile jungle filled with snakes is that most people have little interest in hearing about it. Instead of being interested or curious about my experience, upon hearing I had cancer, most people looked at me as if I had suddenly sprouted horns or a halo. Conversations with old acquaintances were sometimes awkward. Even medical friends seemed to be already writing my epitaph.

During my travels in a wheelchair I had learned to withstand looks of sympathy, pity, and even the Chinese stare that threatened to steal my soul. But I never quite got used to the quickly dropped eyelids and almost imperceptible physical withdrawal that I encountered if I let it slip that I had had cancer.

When I spoke the dreaded "C" word to new acquaintances, or even some of the old, there would be a brief moment of silence. Then, quietly, as if at a viewing, most people would speak with the consoling tones of an undertaker, hastening to tell me about a friend or relative who ". . . had cancer, you know." After a pause, before they considered to whom they were speaking, they usually added, "Oh, how she suffered," adding a brief story of heroism which always seemed to wind down to a depressingly fatal finale.

I never understood why it is thought appropriate to tell someone who has had cancer about the unhappy end to someone else's story. Of course, I had never understood why people, when they found out I had MS, rushed to tell me stories of their troubled son with cerebral palsy, their poor aunt with a stroke, or even their unhappy selves when they broke their ankle. Perhaps, it's an attempt to relate, but most of us with a disability or disease find it easier to relate more to the active and living than to the unhappily disabled or the dead.

In order to survive the misdirected sympathy of such laymen—whether it was pity or pseudo-empathy, I didn't really want to know—I joined a support group. Just as I had sought to join a wheelchair tour when I first started traveling in one myself, I now sought the security of a group of fellow cancer patients and the expertise of a nurse guide.

None of us were quite ready for sainthood, nor, on the other hand, to demonize our disease. We offered each other understanding, encouragement, and reassurance. Best of all, we offered each other laughter and rediscovered that it was indeed the best therapy.

After six months of sharing stories and jokes with my group, as I continued to brave the coils of my doctor's office, I was proclaimed cancer-free. To make sure it was still paralyzed by all the venoms and remained in remission, I returned to the doctor every three months for the next six years, and continued to meet with the group once a week.

Daily tamoxifen, female-hormone-blocking pills, became the machete I kept on my belt to immediately cut off and starve any seedlings that might try to grow. Once started off on the intricate path leading away from the now-dead jungle, I soon was busy getting on with the rest of my life. Now, however, I appreciated the peaks and even the valleys a whole lot more and rejoiced in being alive to experience them.

Even so, during the first five years after diagnosis, I occasionally felt as if I were a circus performer, tiptoeing across a high wire stretched over a deep jungle chasm. I told lots of jokes and prayed that no miasmic cloud would envelop me, causing a final fatal plunge. Ultimately, what saved me was not only the laughter around me but the same thing that kept me going while learning to live with MS. As I proceeded one careful step at a time, travel again became my balancing pole.

* * *

Time now seemed too precious to let fear deter me from adventures I had previously thought too dangerous or too uncomfortable. Soon after finishing chemotherapy, I started with the trip I thought would be the most difficult. A real jungle now seemed much less difficult to deal with than the one I had just escaped, and I was soon on my way down the Amazon River. The way I saw it, I had nothing to lose and everything to gain.

7

Facing up to Fear on the Amazon

Like a warm wet blanket, the smothering humidity in Manaus encased flesh in stickiness, weighted each limb, and caused lungs to struggle for air. Within minutes of our arrival, the clouds, like fluffy pregnant sheep, suddenly broke their water and deluged the city with warm liquid. Although it was like being caught in a waterfall of thin Karo syrup, no one hurried to snap open umbrellas or shrug into ponchos. Any extra burden was unwelcome in this city, where even the touch of the lightest clothing was painful on cutaneous nerves already screaming from the heat.

Many travelers to the Amazon jungle are, if they admit it, afraid of the snakes—long, slithering killers silently waiting to drop from the trees and squeeze one's life away with breath-stealing hugs. Some travelers have more fear of virus-laden mosquitoes, ready and eager to exchange a large dose of death for a tiny taste of blood. But knowing I could take precautions against both, most of my fear was reserved for the heat.

A litany of well-meant advisories from people with MS echoed in my mind. "Heat's dangerous. Heat will cripple you more. We don't do well in the heat," they said. Like me, they had been conditioned to fear it.

Sometimes fear can grab you by the scruff of your neck and shake you to immobility. But sometimes you can break free by hurling yourself into activity. I was not about to give up on lifetime dreams of an Amazon adventure just because I was afraid of the jungle heat.

As we reached our taxi, the rain ended as abruptly as it had started. Less acclimated than the brightly clothed pedestrians drying off quickly in the sun, we raced through the heat in the rattling car, seeking an escape to air-conditioned rooms.

The temperature in the downtown hotel was, however, subject to interpretation. You could choose to consider it as providing some relief from the heat outside or not as cool as had been hoped. Grumbling angrily, small air conditioners struggled to chill recalcitrant molecules of the torpid air. Equally noisy, large ceiling fans labored almost as hard to complete agonizingly slow revolutions. Candles and matches on a desk suggested the situation could be worse. Even so, it was tempting to stay inside for the duration of the trip. But I feared losing out on adventure more than I feared the hellish heat, so the next day we joined the locals whose boats crowded the nearby Rio Negro, the tributary which would soon be joined by another and become the mighty Amazon.

In a big marina near the hotel, hundreds of boats huddled together like impatient coveys of quail. Periodically one would break free, only to be replaced by another eager to join the flock. Jostling amicably at the docks, squat touring boats looked as if they had been created from white, oval balconies—single or stacked in twos or threes—with flat, brightly colored roofs slapped on the top. Competing for space were fat, low-slung fishing boats, slightly lower in the middle than at the ends, which looked like floating jelly beans shaved slightly flat at the top. Scurrying river buses scooted periodically away from the island city like waterbugs urgently rushing to a new home.

The air was filled with the cacophony of blowing horns and the babble of hundreds of voices. T.C. slowly wove my wheelchair in serpentine fashion through the milling crowd. Caught up in the excitement, I almost forgot about the heat. Aboard the boats, sailors tanned to the color of mahogany flashed brilliantly white teeth as they smiled their welcome and their hope. Almost every boat was for hire.

"Hello, hello, come on my boat," said one black haired man with sparkling brown eyes.

"My boat's the best. Why don't you come aboard," said another, face crinkling in a broad smile.

"I take you to the Meeting of the Waters. I take good care of you," said another cheerful man.

Although the river on that late April day in 1993 was nearing its annual forty-foot high, the huge flotation drums under the dock kept it almost level with the main deck of the boat we hired, making it easy to get aboard.

Freeing itself from its jostling sisters at the dock, our boat pulled out to mid-river, where it could chart a straight course. Soon, after slowly motoring a short distance upstream, it passed near a starkly utilitarian waterfront. Here, large seagoing freighters could dock a thousand miles inland to load and unload precious cargo, part of the secret of Manaus' former wealth. But with no big ships presently in harbor, the deserted area looked like a ghost town.

Dotted among the long, narrow wharves, tall, spindly cranes pointed uselessly toward the sky. Near our boat a smaller one beckoned eerily to a dock like a skeletal finger on oil drum knuckles. Just out of its reach, huge cargo containers, interlocked like multicolored Lego pieces, rested like an abandoned nursery school project.

In the distance, a few scattered skyscrapers stood in dark profile against the bright sky. Nearer by, small, brightly hued houses—blue, pink, coral, and white—crowded together in colorful confusion. Jammed together at odd angles on a small terraced hill, they looked haphazardly arranged as if shaken in a giant's hand then hurled in a fit of petulance onto the dirt.

As our riverboat then moved downstream from the city, tree-jumbled river banks showed less and less evidence of man's presence. Only occasional groupings of two or three wooden shacks, either on stilts or on chained-together rafts, were visible in widely scattered clearings. Each of these tiny villages, built on a small piece of family-owned property, had started as a single home then expanded as the family grew.

In small, barely fertile yards snatched from the encroaching jungle, a few scrawny chickens pecked lackadaisically at the dirt, moving slowly in the heat. They ignored the many small children who ran among them to wave at us as we passed. As they jostled for position, the naked toddlers could be seen to have the potbellies of chronic malnutrition. The over-large shorts drooping on their older siblings did not help to hide their painful slenderness. They were conceived, I was told, because there was not much else to do at night without electricity. Birth control, if understood or wanted, was too expensive. It seems a problem wherever people have little education and almost no money. Having no way to solve the problem, all I could do was smile at the kids and wave until they were out of sight.

As I sat on one of the boat's slatted seats, which looked like a park bench and was about as comfortable, I felt as able as any other passenger. With my wheelchair cushion tucked behind my back, I was probably more comfortable than most. As we floated towards the Meeting of the Waters, even the tremendous heat seemed to bother me no more than anyone else. My fear of it, based on inadequate information and experience, was rapidly disappearing.

Feelings of comfort and capability lasted only until I discovered that I would have to transfer to a canoe for a closer exploration of the swamps and sunken forests. On the shallow feeder tributaries, a canoe's shallower draught was necessary to make any progress. We had passed several small canoes carrying one or two waving boys. Sharply pointed at each end, the slender boats seemed little more than hollowed-out slivers of wood.

My vanquished fear of the heat was replaced by a new one. With my poor balance, it seemed impossible to get from the rocking river boat to a bobbing canoe without falling into muddy, piranha-filled water. The other twenty-eight passengers stepped easily into two waiting canoes. T.C. followed them with my wheelchair. As I waited, deprived of the cooling forward motion, the deck of the now-anchored river boat began to heat up so much that staying behind was no longer a tenable option. When T.C. turned to call for me, I knew it was now or never.

Although the guide and crew offered assistance, I was still afraid of the transfer. But after hesitating long enough to become embarrassed, I simply sat down on the boat and swung my legs into our canoe. After sliding on my rear to the canoe's broad rim, I continued until I was sitting on one of the ten cross-boards that served as seats. My moments of fear had been wasted—energy spent for nothing—because, until I actually tried boarding the canoe, I had no real evidence that getting into it would be dangerous.

In the lush, green tangle of the sunken forest, the dark shade made amends for the sizzling heat of the sun. Less comforting was the suddenly oppressive way that the jungle began to close in around the small boats as we went deeper into it. As the river narrowed, an impenetrable mass of vegetation—giant palms, ferns, banana trees, and more than could be named—firmly walled the canoes off from any trace of civilization. Curling, twisting vines and branches seemed to reach greedily for the passengers on the puny boats. In the green, dim light where treetops met over the water, the jungle seemed a hot, malevolent cocoon.

When the motors were turned off so propellers could be

lifted free from the shallow river-bottom, there was a sudden silence. It was instantly followed by a cacophonous din previously unheard over the motors' buzz. Birds cawed and shrieked, almost outmatching the deep profundo bass of thousands upon thousands of frogs. Monkeys chattered noisily as they ran and danced on branches almost too slender to take their weight. Underlying all the sounds, like the insistent drone of a distant bagpipe, was the high-pitched hum of millions of insects.

Despite the presence of twenty-nine other passengers plus two boatmen, a photographer, and a guide, I felt totally isolated from the comfort of human society. A gigantic, dying tree, hollowed and rotted by a leafy parasite, bore mute witness to the law of the jungle that decreed only the fit should survive. Life for many meant death for others. Large fish engulfed smaller fish. Large trees forced smaller ones to die in their shadow, while tiny insects tore the life from five-foot-wide water lilies. Man had already destroyed so many of the ferocious jaguars that once roamed the jungle freely that almost the only place to find one was in a zoo. Survival of the fittest.

I give thanks that, as a disabled person, I am allowed to survive through a combination of both learned skills and the good will of others. By the law of the jungle, I shouldn't be here. Only the fact that society has not degraded to that primitive a level allows me and many like me to exist. If I had been alone in the Amazon, I wouldn't have stayed alive very long. But, neither would most of the other passengers on that expedition. Although I was probably the most vulnerable, mutual cooperation had allowed us as a group to do far more than anyone could have done alone. Such mutual help and cooperation may even be the ultimate answer to the problem of survival on an overcrowded planet, unless of course it's already too late for a people too used to battling each other.

Cooperation was not the word to describe what was happening at the Meeting of the Waters when the river boat arrived there an hour or two later. Where the Rio Negro and Rio Solimoes came together, what was euphemistically called a "meeting" was more like a battle between two powerful forces. After flowing all the way from the Peruvian Andes, both tributaries struggled for dominance for the next several miles. Lively tentacles of lapis-lazuli blue from one poked insistently into the slowly moving mud of the other. Where the fighting was intense, eddies and whirlpools formed, then broke into tumbling multicolored quilts of water—blue surrounding brown or brown surrounding blue.

The two rivers—the "white river," cool and filled with life, and the "black river," warm and almost dead—kept their independence as long as they could before declaring an uneasy truce and merging to become the pale-brown Amazon. As if to celebrate the end of the struggle, pink dolphins leapt high in the air where the Amazon River began.

An hour later the river boat drew near to an island village, home to a rubber plantation. As we slowly moved toward the muddy shore, we passed a wooden dock so decrepit it seemed to be tumbling into the water. Soon the boat pulled alongside a crude structure that was to be our dock. Constructed of a series of double planks, it rested on rickety stilts made of large leafless branches. Every three feet, a long, vertical branch jammed into the riverbed stuck up above the planks as if to keep them from toppling sideways.

The double plank surface of the dock was too narrow for a wheelchair, or even for two people to walk abreast. Either I could walk it myself or I would have to stay on the boat. Now that the boat was stationary, its decks were again becoming hotter and hotter. I faced almost the same decision I had had to make before except here there was more danger of falling. Fear of the afternoon sun competed with fear of the murky water, but the decision was weighted in favor of walking because I didn't want to miss seeing how rubber was harvested. Sometimes fear takes a backseat to stronger desires.

After all the other passengers had disembarked, T.C. carried the wheelchair onto the dock. On his heels, I began a slow walk on the planks, grabbing and holding each vertical support in succession. It was easier than I expected. Again fear had been based on lack of information and experience and had been an unnecessary source of stress. Eager hands grasped my elbows when I let go of the last pole and steadied me on my final step to the ground. Soon I was gratefully sitting in my wheelchair, more comfortable than my companions who had to stand for a half-hour in the shady grove of rubber trees.

As skinny, dark-tanned children focused wide eyes on the group, the guide demonstrated how the once-precious rubber was claimed. Using a long-handled blade shaped like a crescent moon, he made an incision in the tough bark of a tree already heavily striated with cuts. Within seconds, milky-white latex oozed slowly along the diagonal slash to drip into a cup nailed just below the end. Pulled free, the fresh latex stretched three or four feet like a large piece of chewing gum.

Working at night with a kerosene lantern strapped on his head, a plantation worker would start at 3:00 A.M. to place a cut on one side of each of two hundred trees. Once the liquid latex had been collected, he would then build a fire and cook it to make a crude ball of rubber. This was the substance on which rubber barons had made their fortunes before Brazil lost its monopoly on production. Despite very real dangers from jungle predators, or a bad misplacement of the sharp little knife, fear had not stopped the workers who needed their meager wage. Sometimes even reality-based fear must bow to a need to survive.

When we left the plantation and returned to mid-river, a

few young boys in tiny canoes paddled alongside the boat. Competing for attention, one held up a sleepy baby monkey, another a tattered-looking sloth. A third showed off what looked like a tiny, hissing dinosaur, which he grasped firmly by the scruff of its neck. After learning it was a baby cayman, the South American version of the American alligator, I reached over to touch it. Surprisingly, it was warm and dry, not wet and slimy as I had expected. But with its soft scaly skin, the cayman felt like a reptile, a class of creatures for which my dislike was so strong it was hard to distinguish from fear. Quickly withdrawing my hand, I didn't yet know that I would soon be dealing with a much larger one.

* * *

After returning to Manaus to rest for a day, we traveled by boat to a remote village where a scattering of non-air-conditioned visitor's bungalows would for the next week be a relatively luxurious base from which to make deeper forays into the jungle. The only way to get up the slight hill to our bungalow was to climb a narrow, winding path. Roughly underlaid with roots, it was almost too uneven for my wheelchair to traverse. T.C. had brought along what looked like an extra-wide seat for a child's swing, hoping to use it for carrying me in places too difficult to walk. He explained proudly that this was a Hoyer sling, the best to be had, but the first time we used it was the last.

After inserting two poles through loops at both ends of the canvas, T.C. and our guide hoisted the poles to lift me in the rough canvas sling. My back began to burn as I started sliding toward the guide. As he tried to lift his end level with T.C.'s, his straining muscles stood out like fibrous cords under his thin skin. But, a full foot shorter and more than a hundred pounds lighter than T.C., there was no chance of an even match.

The further they walked, the deeper the sling dug into my back as I slid closer and closer toward the guide. Sweating heavily, he struggled up the treacherous path but gradually started to wilt. When his knees began trembling so hard they almost knocked together, he nearly lost his grip on his end of the pole. As he jerked it up, my abraded back screamed for relief. I called the men to a halt and decided to climb the rest of the way myself, with T.C.'s assistance. The choice was not determined by fear of falling from the sling, but because of immediate pain, which is sometimes a more powerful incentive.

When I reached the bungalow, I wondered how anyone could make it through even one night in this dim, stifling room. Tiny, mosquito-netted windows barely offered a breeze, but the door could not be kept open because deadly mosquitoes continually foraged for blood in the jungle, trading for it malaria by night and yellow fever by day. In the tiny room, the night heat sucked away energy like a starving vampire. But despite the warnings from folks with MS or those who thought they knew something about it, I found in reality that I was no more enervated than T.C. The assumption that MS made me different than anyone else proved, at least in this case, to be false.

Only the quiet geckos, who scampered about the walls after stuporous insects, seemed unaffected by the torpid heat. Even scorpions and giant cockroaches, kept at bay by bedposts carefully placed in cans of water, seemed to scuttle in slow-motion away from the light stealing in through the net-covered window. The only thing to do was to work on being comfortable and having fun until the boat came back to take us away. Replacement of fear by pragmatism often makes a difficult situation tolerable.

The jungle near the bungalow was too dense to penetrate in a wheelchair without the vigorous wielding of a machete but, although unable to enter it, I was nearly entrapped by it anyway. Wanting the independence of a real explorer, I decided to wheel myself along a narrow dirt path cleared along its edge. However, while gazing into the jungle's congested, green depths, I failed to keep track of the small front wheels nearly hidden by my legs. With a sudden crunch, one dug into the mulch on the edge of the hard-packed path, and my wheelchair crashed over sideways.

Unhurt except in pride, I struggled to get free, but remained trapped in the rigid, metallic embrace of its armrests until an alarmed T.C. pulled it away. Humiliation was quickly replaced by a vow to practice my new-found independence in a safer place. But after the taste of freedom, there was no going back to my former fear of self-propulsion.

While I was living in the village, the only real respite from the jungle heat was offered by traveling on the water. By day, sightseeing journeys through swamps and sunken forests alternated with three-hour fishing trips. At night we hunted cayman in the swamps.

On the first night, only the slim beam of light from our guide's powerful flashlight fashioned a trail into the darkness for our small, motor-powered boat. Mottled green and black areas at the water's edge seemed to move in the light as tiny invisible predators hustled about in search of a meal. Like a living thing, the shore seemed patiently waiting to entrap our small boat if we carelessly came too near. Suddenly the boat made a shallow sweep directly into the swamp.

Two small red eyes, transfixed in our flashlight's powerful beam, glared at us out of the jungle. Using them as a beacon, the boatman motored slowly towards the land, then suddenly cut the motor and quietly lifted the propeller out of the water. The guide pressed his finger to his lips. As we drifted closer to where the water ended and the land began, the boatman cautiously used a long pole to direct the boat.

As we came closer and closer to the unblinking red eyes, the guide began to whisper instructions. Handing me a long bamboo pole with a small hangman's noose affixed to the end, he instructed me in how to use it as a snare. His voice was barely a puff of breath on my cheek as he explained that he would grab the cayman the instant it was caught.

As we drew silently towards the shore, I flattened myself from head to mid-thigh across the aluminum shell of the prow, taking care to keep the noose out of the water. Closer and closer we came to the eyes. Suddenly they disappeared. We were almost directly on top of our prey, which remained immobile in the shallows. I didn't dare breathe.

"Do you see it? Do you see it?" whispered the guide very softly, his lips almost touching my ear.

I moved only my eyes to search the darkness next to the boat. A very slight creak betrayed the guide's change of position as I peered intently at the mottled swampland. Then, a move so slight as to be only a suggestion gave away the cayman. Almost perfectly camouflaged, he lay very still. I nodded my discovery to the guide.

Very slowly I moved the noose just above and ahead of the quiet creature and lowered it nearer and nearer. I paused to take a deep, silent breath. Not a sound came from anyone on the boat. Tensed for action, I suddenly dropped the noose millimeters ahead of the light-dappled nose.

In a split second the cayman became aware of something near and sprang into action. Terrified, it ran frantically forward—straight into the noose. I jerked sharply up, pulling the hangman's knot tight. With no time for thought, I threw myself upright and back to land on my knees. Fighting to keep the noose closed, I tried to haul the thrashing creature up toward the boat. The line started to cut into my hand as its struggles almost pulled it out of my grasp.

With a movement so fast that his hand was a blur, the guide quickly grasped the thick neck behind the snapping jaws. As he lifted the struggling cayman into the boat, it could be seen that this was no placid pet in a little boy's canoe. It was instead an angry five-foot long adolescent. As it whipped its tail about, I threw myself out of the way and onto a seat. Then, holding the suddenly quiescent creature, the guide slowly sat down next to me.

Tightly grasping the cayman's neck with one hand, he used the other to lift its bloated belly. The long tail fell straight across my lap. I sat rigid, then relaxed enough to smile for a picture. Looking away from its blunted nose—more pointed than an alligator's, less pointed than a crocodile's—I tried not to think of the sharp rows of teeth hidden in the grimly clenched jaw. The vertical slits on the creature's unblinking eyes made it appear to be making evil plans as it glared, unseeing, straight ahead.

Suddenly the great tail slashed across my waist, breaking free of my loose grip. Alarmed, the guide leapt to his feet as the creature began to thrash violently back and forth.

"You have enough pictures?" he asked with a weak grin, holding the creature as far away from himself as he could.

Unable to take my eyes off of the cayman's snapping jaws, I nodded.

After he hurled it off the boat, it stayed perfectly still for an instant, then with a great slash of its tail it was gone.

We went out three more times, but I decided that one good catch was enough. T.C. caught an even bigger cayman but, to my relief, the guide said it was far too big to bring aboard the boat.

Piranha-fishing in quiet, shady coves offered more relaxing sport by day. We announced the availability of food to the sharp-toothed predators by noisily beating the water with the tips of long, bamboo fishing poles. Theoretically, they would think a large, thrashing meal had just fallen into the water as our hooks bearing chunks of raw meat floated down from the murky surface.

If a nibbler was not immediately hooked after its tiny tug on the line was felt, all that came to the surface when the line was brought up was a bait-less barb, shining darkly empty at the end of the line. Many times, not even a warning nibble could be felt before a check of the hook revealed it to be completely stripped of meat.

As if stuck in glue as thick as the oppressive humidity, the hands of time moved slowly as we all sat and fished—the boatman, the guide, T.C., and me. Developing a simple rhythm, I beat the water, dropped the hook, brought it up, replaced the meat—then started all over again. The soporific heat and lulling harmonies of insects and frogs acted as a powerful sedative. Despite the activity, my eyelids began to droop.

Suddenly I had a firm hit and the tip of the pole dipped sharply. Instantly wide-awake, I yanked my pole to vertical as the small, round-bellied piranha fought like a ten-pound bass. Red and yellow flashes of light gleamed off its shimmering scales as it hurled itself about in desperate attempts to dislodge the barb. The grinning guide quickly dropped his pole and deftly fielded the struggling fish with a long-handled net. As it quieted, he allowed me to dangle it for a photo before quickly grasping it behind the gills. Then, with one quick motion, he pulled the extra-long shank of the hook free of the still snapping mouth, then tossed the fish, weakly flopping, into a large plastic bucket.

After two or three hours of fishing, the bucket was nearly full. As I leaned near to take a closer look at the catch, the guide lifted a warning hand. Pulling a stalk of celery from a soda cooler, he touched it gently to the pointed mouth of one fish whose body had already started whitening with death.

Snap! The still jaws suddenly opened, then viciously closed. Snap! Snap! The sound reverberated around the bucket as the jaws of others abruptly closed in dangerous spasms. Each tiny post-mortem reflex could have easily taken off my finger. Even after death, these voracious eating machines, able to strip a plump cow down to bone in under twenty minutes, still sought a victim.

Hungry piranhas were far from my mind, however, during our last half-day at the village. An initial plan to lie across a wide, colorful hammock began to lose appeal as the sun rose higher and higher in the sky. Soon T.C. rolled me to a narrow dock where more than a dozen small canoes bumped lazily together. He soon had permission to take one out.

By now, the seated method of entering a canoe had become an easy skill, and soon T.C. was paddling us upriver. Formerly cacophonous, the sounds of the jungle now seemed comfortingly melodious, and branches over the water no longer seemed to avidly grasp for a victim. Fear quite often disappears when familiarity takes its place.

On our return downriver, unable to remember all the forks we had taken, I became afraid of getting lost. But such fears, based on mistrust, were revealed as unwarranted as T.C. unerringly made the right choices. Near the village, we approached a small island, where three monkeys sat intently watching our progress. Knowing that an expanse of water was to a monkey as much of a barrier as the bars of a cage, the locals had cut a channel across a short peninsula to keep them as pets.

As T.C. smoothly rowed into a shady cove, I found a few pieces of dried-up apple dropped by those who brought yesterday's fruit to feed the village pets. Instantly, the monkeys went into motion. When little water was visible between the land and the canoe, the male jumped on board. Quickly grabbing the piece of fruit, he held it up like a trophy and jumped back on shore. When I coaxingly held up another piece, the whole family came into the canoe—the dad, mom, and the baby who rode on her back. Watching carefully for any attempt to bite, I placed small segments of apple in the monkeys' tiny, outstretched, black hands, which grasped them ever so gently before conveying them to eager mouths.

Again the male jumped back to shore, but the female and baby stayed. Soon the mother was on my shoulder, gently picking through my hair while her baby snuggled in my arms. As the watchful male hung from his tail on a nearby tree, T.C. batted his hands to send him flying backward. On the return swing, the monkey extended his hands to slap T.C.'s and continue the game. Although he seemed willing to play indefinitely, the male quickly jumped on board when I found two small blackened bananas I had previously missed.

After the fruit was gone, T.C. and I became buried under the busily investigating trio as they leapt to perch on empty laps, shoulders, and heads. When it was time to go, the monkeys did not want to leave. Only a vigorous splash of water on their fur convinced them to get out of the canoe. After my return home, a friend seemed shocked that I had played with semi-wild monkeys.

"You could have caught rabies!" she exclaimed.

At the time, it had never entered my mind. One person's fear may be another's great adventure.

We returned to the village just in time to catch the biweekly river boat for the trip back to Manaus. That afternoon, it could find no berth amidst the covey of boats already moored at docks outside the city. After our boat anchored past the last dock, the crew threw out a gangplank barely a yard wide to span the thirty-foot distance to shore. Uncomprehending at first, my heart sank when I saw fellow passengers form a single file and start to walk onto it.

Hoping an alternative would present itself, I turned my attention to a nearby building. Although widely spaced supports under its corrugated tin roof left it open to the muggy air, the interior was dim. Inadvertently I met the glassy eyes of a man with disheveled black hair who stood just inside. His peg-thin legs seemed barely able to support him as he leaned against the sagging plywood counter of the bar. As I stared at him, his eyes suddenly focused on me. A weak grin curled his mouth like a sneer as he toasted me with a partially filled glass.

Nearly spilling in his shaky grip, the fluid in the smeared glass may have been cheap rum or beer. But for this man, its contents were more likely liquid death. As I shuddered and glanced away, I looked back toward the water. A light, iridescent sheen of oil barely covered its black stagnance. Remembering the piranhas, I wondered if I was looking at my own liquid death.

Finally the moment of truth was at hand. Except for the crew, who were getting ready to leave, I was the only one left on the boat. Hope died for finding another way of getting to shore. I hesitated, but no other choice was presented. Fear swept over me. Even T.C. had been careful of his balance as he carried the awkward burden of my wheelchair. But as I looked at my fellow passengers, laughing and chatting on the shore, I was as afraid of being left behind as I was of walking the gangplank.

As the guides on the shore and the crew on the ship called encouragement, I tentatively took my first step onto the plank. Not daring to look at the water nor at the distance ahead, I concentrated on watching my feet. Time seemed to stand still as my sneakered feet and that splintery, sun-grayed board became my entire world.

Intense concentration kept fear at bay until I flicked an inadvertent glance at the water. With shaking knees, I came to a halt, not daring to move. I stood for a minute trying to decide if I should sit down and slide the rest of the way on my rear. But large splinters protruding from the wood made this an unattractive option. Finally, chanting a soft monologue of encouragement, I willed myself to take one step, then one more ... and yet one more.

Suddenly my heart began pounding as I started to fall. But it was only because eager hands had grasped my elbows to pull me the last couple of feet to shore. As if a television set had been suddenly turned on, I became conscious of many congratulatory voices and an abundance of broad smiles. I had made it. I was safe. Even though I had done no more than anyone else, I felt proud.

In the hot crucible of the Amazon river basin, confrontations with fear had been hard to avoid. When fears are based on reality, one doesn't usually have the luxury of dwelling on them; it is too necessary to use one's energy for survival. In the Amazon, however, my survival was never really in question. When not everyone seemed to share my fears, I took a better look at them. Once carefully examined, most of them had been only F.E.A.R, False Evidence Appearing Real, and merely a waste of energy.

Many children who fear monsters under their beds at night can be reassured by simply shining a flashlight underneath. As adults we can similarly learn to face many of our fears by shining the bright light of reality on what we think is evidence supporting them. When we discover that the evidence is false, as it usually is, like children we can learn to put fear away and wake up to new vigor.

* * *

Unfortunately, for the hapless people of Cambodia, the bright light of reality only accentuated their country's disaster in the late 1970's. Even in 1993, the evidence of it was still brutally real.

8

A Beggar's Eyes: Cambodia's Lesson

In the sprawling heart of Phnom Penh, a beggar-woman bent low by age slowly worked her way toward our hired car where it stopped for a signal. As she hobbled closer, trying to peer in our windows, her soft mumbles grew more intense. Barely lifting her head, she pressed the deformed, arthritic fingertips of one hand together and touched her wrinkled mouth in the universal signal for food. Inadvertently I met her gaze, then quickly looked away. So dull that even bright sunlight brought no sparkle to their depths, her ebony eyes were like dead coals in the burned-out wreck of her face. Totally devoid of pleading or hope, they contained no evident spark of humanity.

"Never look into a beggar's eyes," I told myself uneasily. "Never look into a beggar's eyes."

Perhaps most would look away when confronted with someone who brings deeply buried fears to the surface. No one wants to believe that one can be so vulnerable to misfortune. But, thinking of all the times I had also been a recipient of someone's quickly averted glance, it seemed wrong that I had done the same to another. When I looked back, however, it was too late. Head down, she was already shuffling away.

As our car left the intersection and continued down the wide avenue, the feelings of guilt, unleashed by my initial self-admonition never to meet a beggar's eyes, soon faded. There were many eyes much more pleasant to meet. Pedestrians rushing about in dark business suits despite the muggy heat glanced over curiously as we went by. People even smiled and waved as they looked down from the long balconies that decoratively striped the low-rise buildings.

At first, the streets of Phnom Penh seemed like those in any other large Third World city as a crush of vehicles—bicycles, motorcycles, pedicabs and cars—challenged each other for space on the wide avenues. But on looking more closely at the scene, I realized something was missing. On both streets and sidewalks, there appeared to be almost no people of middle age. This fleeting impression was soon lost, however, crowded out by others when we reached our hotel.

As T.C. pushed me along the damp grass near the entrance, the placid brown eyes of a water buffalo, almost level with my own, gazed at me with liquid softness. Soon, dull curiosity satisfied, it lazily lowered its head to feed. Such calm interest was not the reaction of the desk clerk, however, when he noticed my wheelchair. Eyes filling with concern, he rushed around the desk to inform us that our rooms were on the second floor, but the only elevator was broken. As he apologetically averted his gaze, the slightly built young man in the well-fitting, blue business suit explained that replacement parts were not due for two weeks.

After he made a flurry of phone calls, the worry in his eyes was finally replaced by relief. Smiling broadly, he directed us to the narrow dirt path that would bring us to a nearby hotel with more accessible rooms. Alert for snakes in the thick undergrowth at the edges of the path—as T.C. carefully pushed me along the strip of packed earth—I forgot about Phnom Penh's missing generations.

Thoughts about them returned with magnified force, however, the next day at Tuol Sleng Prison. Now a museum, this was where the Khmer Rouge, the Communist "Red Cambodians," detained and interrogated people between 1975 and 1979. Surrounded by well-tended lawns and a scattering of palm trees, the white-painted building at first seemed as graciously welcoming as a Southern mansion. Low-walled outer hallways, sectioned by tall pillars, were wrapped around each of its three floors like old-fashioned verandas. On closer inspection, however, a net of barbed wire over all the outer openings gave an ominous hint of what was inside. A small sign noted: THE BRAID OF BARBED WIRES PREVENTS THE DESPERATE VICTIMS FROM COMMITTING SUICIDE.

Our guide, an educated, thirty-five-year-old Khmer woman, an ethnic Cambodian who had escaped her country just in time, explained another sign, which contained a list of questions: "Do you know a foreign language? Have you ever been out of the country? Do you wear spectacles?" With world-weary eyes more typical in one fifty years her senior, she explained that these innocuous-appearing questions had been used, along with others, to identify those who knew anything of the outside world or of their own history and culture. One by one, those who possessed knowledge—teachers, diplomats, dancers, and monks; even those who merely used glasses to read—had been pronounced guilty of a capital crime and killed. The sign may as well have read: "Abandon hope all ye who enter here." A place of exquisite tortures and gruesome deaths, the prison had been a hell worse than Dante's inferno.

Looking both angry and deeply sad, the guide explained that the goal of the regime headed by Pol Pot had been to set the clock back for a full millennium. By restarting history at "Year Zero," those with military knowledge could easily control an impoverished group of uneducated peasants. Anyone judged unwilling or unable to become such a simple peasant was put to death.

At the entrance to the prison, a large map of Cambodia immediately commanded attention. Against a baby blue background, the striking bas relief was composed of lustrous bronze, ivory, and ebony human skulls. Carefully arranged among them, long slender bones marked provincial boundaries, while a few red-painted plywood rivers connected to a crimson-hued lake looked like rivulets of blood streaming into a coagulating pool.

After T.C. went to explore and photograph the cells upstairs, the guide stayed with me on the first floor, where the terrible purpose of this "security prison" was revealed in horrifying detail. Prominently displayed in the first few cells were instruments used for torture—now, only rusting mementos of madness. Paintings above each clearly illustrated its use. How many eyes had looked with terror at these cruel objects? Who had dared recall the details of torture enough to capture them in paint?

In one white-washed cell stood a heavy, darkly stained wooden table with manacles attached at each end. Pictured above was a blindfolded man lying on it with body arched back like a bow because wrists and ankles were lashed together by thick chains. Mouth open in a silent scream, his forehead furrowed in agony as a boy soldier grasped his hair and pulled his head up to slit his throat. Bright-red drops of blood were already flooding into a pool on the table.

Paintings in cell after cell gave testimony to man's grim creativity in causing pain to others. Worse than the rooms with the graphic paintings were the cells where evidence of human presence made the past horrors more real: a frayed prison uniform, neatly folded, lying at the end of a narrow iron cot; a thin, blood-soaked pallet lying on the hard cement floor. But the very worst was yet to come.

Immediately upon entering a series of mass detention cells, I was assaulted by stares from hundreds of eyes. Every wall was filled from floor to ceiling with small photos of men and women and even of a few toddlers clinging to their mother's skirts. Scattered among the hundreds of images of Khmer people were those of a few foreigners—American, French, Australian—

caught in the wrong place at the wrong time. All wore numbered discs for identification. Like the Nazis, the Khmer Rouge had kept meticulous records of their victims, including photographs taken before and after their death. After combining a guess of the ages of those in the photos with a quick calculation of dates, it seemed I had found the awful reason for the absence of middle generations among the people of Phnom Penh.

Many eyes in the photographs were resigned. Some were hostile or hopeless. One young man's eyes crinkled with a smile as he posed for the camera. Perhaps he had hoped a cheerful manner might save him, or maybe it was just his nature. But neither had done him any good.

Some blackened eyes were squinted by unimaginable pain. Some, peering from swollen faces above blood-encrusted lips, looked dazed. A few half-closed by death seemed to jeer at their tormenter. Although I didn't want to meet these eyes, it was difficult to turn away. Surrounding me on all sides, they demanded acknowledgment.

It was almost easier to glance at the clinically graphic, full-body photos of those whose eyes had been closed by death. Only massive bloodstains on a cloth beneath one man's head and the dark holes stitched raggedly across the back of another gave away the fact that they were not merely sleeping. As many as one hundred people a day had met their fate here when this prison was at it peak, and there were many other similar prisons. In four short years, over two million people had died in Cambodia—one out of every four—not only from torture but also from forced labor and starvation.

After seeing this sad, grim gallery, any claim I had ever made to worry or pain seemed merely an affectation. Not only had my life never come even close to being this terrible, but I had also been given many chances and choices. Not the people portrayed on these walls. For them there had been no such hope... and they had known it.

To add insult to injury, a final wall held photographs of cheerfully smiling young prison guards with arms wrapped around each other's shoulders. In one, Pol Pot himself, head broker of all this misery, stood laughing with his men. As I looked directly into the smiling face of evil, I knew that I could never again doubt its existence.

"What happened to Pol Pot?" I asked the guide.

"He retired to Paris with his new wife. It is said he lives in luxury, but he's not really retired. He is still the head of the Khmer Rouge. Gangs of armed teenagers under his control roam the countryside to make trouble. They plant mines, rape the women, and march the men off for hard labor. If villagers hear they are coming, most abandon their homes and run for their lives. We are a nation of refugees."

"This is happening now?" I asked, unable to comprehend that Pol Pot's evil was not only unpunished but was still going on.

"Yes," said our guide, looking almost as stunned as I by her country's history. "We are grateful that your U.N. has come to help keep the peace. With their help we hope to have our elections soon. But then they will leave and nobody knows what will happen."

Having seen more than enough, I rolled to the outside hall to wait for T.C. The guide insisted on helping. "What happened here, it's... it's... " I couldn't find the words.

"I know," she said quietly. Then, her tone becoming urgent, she said, "Please, when you go home, remember us." She moved to look deep into my eyes before she continued: "Your countrymen say, 'Remember the Holocaust, so it doesn't happen again.' But it did happen again! Here. It was only fourteen years ago, not fifty. I think maybe it is even happening right now in Somalia, and perhaps Bosnia. Please tell America about Cambodia.

About what happened to our people." Still gazing steadily into my eyes, she softly added, "Please."

Although wishing to forget what I had seen, I didn't know how anyone could. I promised, but I wondered if I would ever get anyone to listen. Living in a country that still refuses to fully acknowledge its disabled, I doubted many would want to be confronted with a horror as great as this. But I promised, and meant to keep my word.

* * *

The enormity of what had happened in Cambodia was only revealed fully when we traveled to a place in the countryside nine miles outside Phnom Penh. After a peaceful drive through a seemingly endless green patchwork quilt of rice paddies, we arrived at the extermination camp of Choeung Ek: the "Killing Fields." Those who had not died and been buried at the prison had been brought out here in an attempt to keep their fate secret.

Lush, green, fruit trees circled a large dirt clearing with a tall, white memorial at one end. Several large, square pits nearly hidden by weeds were scattered around the distant periphery. Since the grassy approach to them was slicked by dampness and made a treacherous surface for a wheelchair, I encouraged T.C. to see them with the guide while I explored the smooth paths closer to the center of the clearing.

As I rolled toward one large sign, a few French visitors appeared to keep a careful distance from each other as they walked quietly away.

"*Bon jour*," said one softly when he noticed me. "*Bon jour*," was the quiet greeting of another.

"Hello. *Bon jour*," I said in return. But none would meet my eyes.

The sign contained a crude map of the site with small

squares to represent the mass graves that had been discovered. After reading a few of the abrupt sentences that described the contents of some, I thought I understood the odd reserve of the French. One grave had been filled with men's skeletons. Another, with those of women. Yet another had been found to contain only men's skulls. When I noticed a smaller square labeled "skulls of babies," I started to cry.

Hoping to find solace by visiting the tall stupa, a Buddhist memorial to the dead, I attempted an off-path shortcut across the clearing. Suddenly one of my front wheels jammed against something on the ground. Half-buried in the light-colored dirt, a human rib was wedged under it. About two feet to my left, a small fragment of vertebra was working its way through the earth, and just ahead of me was a long, slender piece of shattered armbone. In horror, I realized I was rolling directly on top of one of the unopened graves. Glancing down as little as possible, I hurriedly rolled the rest of the way to the stupa.

It would have been hard to be unaffected by the mute evidence all around that the veneer of civilization is as easily broken as the veneer of health or of any other illusion we hold dear. The Cambodian people, before the four years of genocide, like a healthy person before physical ability is taken away, undoubtedly thought they would live a normal life—working, raising families, and growing old. But, unlike a person who becomes disabled or ill, they had no friends, allies, or institutions to help them. Such a situation was difficult to imagine, until a large sign in front of the memorial brought it cruelly to life. It read:

> THE MOST TRAGIC THING IS THAT EVEN IN THIS 20TH CENTURY ON KAMPUCHEAN SOIL, THE CLIQUE OF POL POT CRIMINALS HAD COMMITTED A HEINOUS GENOCIDAL ACT, THEY MASSACRED THE POPULATION WITH ATROCITY IN A LARGE SCALE. IT WAS MORE CRUEL THAN THE GENOCIDAL ACT COMMITTED BY THE HITLER FASCISTS, WHICH THE WORLD HAS NEVER MET.

With the commemorative stupa in front of us, we imagine that we are hearing the grievous voices of the victims who were beaten by Pol Pot with canes, bamboo stumps or heads of hoes, who were stabbed with knives or swords. We seem to be looking at the horrifying scenes and the panic-stricken faces of the people who were dying of starvation, forced labour or torture without mercy upon the skinny body, they died without giving the last words to their kith and kin.

How hurtful those victims were . . . before their last breath went out. How bitter they were when seeing their beloved children, wives, husbands, brothers or sisters were seized and tightly bound before being taken to the mass grave while waiting for their turn to come and share the same lot.

The method of massacre which Pol Pot clique was carried upon the innocent people of Kampuchea cannot be described fully and clearly in words because the invention of this killing method was strangely cruel, so it is difficult for us to determine who they are for they have the human form but their hearts are demon's hearts.

They have got the Khmer face but their activities are purely reactionary. They wanted to transform Khmer people into a group of persons without reason or a group who knew and understood nothing, who always bend their heads to carry out Angkor's orders blindly.

They had educated and transformed young people and the adolescent whose hearts are pure gentle and honest into hideous executioners who dared to kill the innocent and even their own parents, relatives and friends.

They had burnt the market place, abolished monetary system, eliminated bodies of rules and principles of national culture, destroyed schools, hospitals, pagodas, and beautiful memorials such as Angkor Wat temple, which is the source of pure national pride and bears the genes, knowledge, and intelligence of our nation.

They have the human form but their hearts are demon's hearts. It seemed the only plausible explanation.

Beyond the sign, the gracefully proportioned stupa towered more than ten stories above the hard-packed ground. Appearing dreamily surrealistic, small puffy clouds drifted across its shining windows, gentle mirrors of the sky. Upon looking through the glass, however, the impression of beauty quickly vanished. From bottom to top, it was filled with more than eight thousand skulls. Organized by sex and age, they were carefully packed on multiple tiers of shelves. With their empty sockets staring blindly in all directions, their hollow gaze was difficult to escape. When T.C. joined me we quickly glanced into each other's eyes, then, like the French visitors, we kept our distance and didn't speak.

* * *

Phnom Penh, on our return, seemed to be covered by a gauzy pall of death. Dimming the sun, it barely covered the quiet body of a city still embalmed by residual fear and grief. Despite the stunned hopelessness in the faces of many people, there was still some evidence of a will to overcome their tragic past.

While waiting the next day for a flight to Siem Reap, I was startled by the innocent delight in the expression of a young saffron-robed monk who sat next to me in the airport. Dark and sparkling, his eyes betrayed an inner excitement he could barely contain.

Smiling shyly, but eagerly, he carefully said, "Hello. Where are you from?"

"America," I answered, then stopped, unsure of the protocol for proceeding when in the presence of a young Buddhist monk.

He nodded and smiled even more broadly than before.

Seeming to struggle for more words, he spoke in rapid Khmer to the man in a white shirt who stood next to him.

In response to the young man's request, the older man explained that he was escorting his friend to the monastery at Angkor Wat, an ancient temple near Siem Reap. This was to be his first time on a plane and first time away from Phnom Penh. As the monk listened intently, his friend told me that I was the first American the monk had ever met and he was eager to practice his English.

"Tell him that he is the first monk I have ever met," I said.

Laughing, the older man turned to the younger with a translation. The monk's smile, if possible, broadened even further but he sobered a little as he pointed at my wheelchair. "How?" he asked, searching for more words but failing.

"Accident," I said, slapping my palms together with a slight crash. It was an easier explanation than trying to describe MS in tropical countries where it is generally unknown.

Both men nodded wisely. There was a moment of respectful silence before the monk haltingly asked if I had any children, a seemingly obligatory question in the Third World. Interested inquiries about country, family, and, in my case, cause of disability appear to be necessary to define a foreigner's identity. When I said I had no children, the men both looked sad but did not pursue the topic. As usual, no one asked if I was married. Perhaps in the Third World any mature woman is presumed to be a wife.

After this basic exchange of information, we struggled to find a topic of conversation. Despite lots of smiles, we had little success. In addition to the language barrier, we had too little commonality of experience to sustain a discussion. When the flight was called for early boarding, we wished each other well, then, almost bursting with excitement, the monk headed rapidly for the gate. I didn't have the heart to explain that only boarding for the disabled had been announced.

* * *

From Siem Reap, we were taken to the temple complex at Angkor. Well-constructed of stone, these temples and royal buildings were all that remained of this ancient capital of the huge Khmer empire that at one time covered most of Indochina. The straight dirt road from Siem Reap to the monument ran like a flat, brown ribbon through the dense jungle. Occasionally there were breaks through the tangled web of trees and high grass where narrow paths, flattened into existence by walking villagers, joined this main road.

Looking out of place, large red signs on red-and-white striped poles dotted some paths like randomly strewn confetti. A large, white outline of a skull and cross-bones was sandwiched between huge warnings at top and bottom which were written in both English and Khmer. Almost screaming for attention the warnings said: DANGER!! MINES!!

On these paths, men walked slowly as they carefully manipulated long, metal poles to sweep flat, silver saucers back and forth near the ground. Cambodia was not, our guide informed us, a country in which to wander off paths clearly marked for tourists. Only after returning home did I find out from a souvenir newspaper that the Khmer Rouge had been active not far away from Angkor Wat on the day before we arrived. What we had seen had not been merely a routine search for old mines which crowded the countryside. The intense activity was directed at detecting any new mines laid the night before.

Only a few hundred feet down the road from the busy groups of men, a white two-story building sat in splendid isolation in a clearing. On the low, outer wall of its open first story, a bright-red skull and cross-bones had been stenciled on each side of the doorway. Large, ornate black letters further to each side proclaimed the small building as the MINE FIELD BAR. Perhaps this was where laughing eyes would triumphantly meet above raised glasses after one more day of survival or where glassy ones would instead stare deep into the amber liquid at the bottom of a glass at the thought of what could have happened.

As the morning heat increased, our driver-guide stopped at a large clearing where the availability of Pepsi and Fanta was announced on signs atop a cluster of four or five wooden stands. As usual, I was quickly surrounded by children, but, unlike other times, their interest seemed less in my wheelchair than in having a captive audience. With a hint of near-desperation in their eyes, they urged me to buy small handicrafts. As they attempted to press them into my hands, their threadbare clothes—long skirts for the girls, droopy shorts for the boys—could not hide the skinny limbs and swollen bellies, the stigmata of chronic malnutrition. Many infants, riding the nonexistent hips of sisters barely two or three years older, displayed oddly reddish hair, the signature of protein starvation. Though it would probably have little effect on their impoverished lives, I could not refuse them several sales.

While I made my selections, T.C. tried to amuse them with a finger game. A hit with children around the world, it was one where you put one hand behind the other, then fold one finger back to make it appear missing when the hand is brought back in sight. When the motions are reversed and the finger is "magically" restored, little children laugh and older ones want to know how the trick is done. But not here. Not these serious children. With no expression in their dark eyes, they stood silent when the trick was completed.

One particularly sober-faced little girl hung back slightly behind the others, cradling one arm close to her body. When I scrutinized her more carefully, I realized it had been crudely amputated at the elbow. With wide eyes, she stared unblinkingly at T.C.'s fingers. Suddenly it was clear why the game did not amuse these children. In a country where mines were a daily danger, especially to those who wander off the paths, they already knew at too young an age that there was no magic way to restore missing body parts.

Realizing the problem, T.C. quickly switched to blowing up balloons. Soon the children, eyes wide with yearning, were giggling delightedly as they waited to get one. A flash of color on the periphery of the group caught my attention. A tiny old woman with head loosely wrapped by a bulky fuschia scarf was watching the group intently. Wearing a citron-yellow blouse, ornamented with gleaming circles of tin and paired with a screamingly bright purple-and-white skirt, she was a gaudy ragbag of color as she edged near the laughing children.

When I caught her eye she stopped like a frightened bird, so I looked away. Each time I glanced back, she was closer. Black eyes sparkling with interest, she soon worked her way to the middle of the group. Soon her tight-lipped mouth began to curve into a smile. When T.C. smiled at her and offered a balloon, she eyed it wistfully but shook her head and backed slightly away. Only after all the children had been given theirs would she finally accept one. As she gave T.C. a happy, gap-toothed smile, years of age and worry seemed to drop from her face. Watching her, it seemed a shame that balloons and the happiness they bring don't last.

* * *

Cambodia seemed a place where people would always be aware of impermanence. If the Khmer Rouge didn't drag the people back to a primaeval age, the jungle and the monsoons ultimately would. This was starkly evident at Ta Prohm, an ancient temple unearthed when the "lost city" of Angkor was rediscovered a hundred years ago. Left as it had been found, the destruction produced both in the remote and recent pasts was evident even at a distance. Each tall, phallus-shaped spire looked as if God-hating soldiers had tried to emasculate the power it represented by roughly slicing off its vital tip. Yet, closer inspection revealed that the jungle had begun an aggressive effort to reclaim the temple long before the Khmer Rouge mutilated it and had stubbornly continued its assault long after they left the holy place.

Carefully staying within the cordoned-off area, T.C. and I followed a narrow dirt path among the huge, tumbling structures. Fist-thick vines grew into the mortar between the blocks of a few intact archways, while huge roots were undermining and uptilting their bases. Strewn on the ground like a child's discarded toys, random clusters of broken sandstone cubes revealed the ultimate result of such intrusion. Whole cloisters of this twelfth-century monastery had caved in, leaving it almost impenetrable by even the most able explorer. It seemed as if Olympus had fallen from the sky and the jungle had been working ever since to hide it.

Feeling like Indiana Jones in the Temple of Doom, I moved from my wheelchair to sit on one huge, moss-covered lintel near a delicate carving almost covered by a network of tendrils. As I tried to push them away, light rustlings betrayed the presence of small creatures moving amidst the wreckage—lizards, frogs, and perhaps giant spiders. I wondered if a small Hanuman snake—light green and deadly—might even now be quietly slithering too close for comfort. In the dim, greenish light of this hidden place, I would be unlikely to notice it before it was too late. Suddenly, a bat flew silently out from somewhere in the dark, inner depths. As a chill brushed my spine, I noticed that not even a bird dared to sing. Soon I headed for the security of my wheelchair and rejoined T.C. and the guide, ready to head for monuments that had been reclaimed from the jungle, at least for now.

There seemed to be more hope for the future in the

reclaimed city of Angkor Thom. On the towering Face Tower, the smooth and seraphic bas-relief faces seemed to be trying to break free from their imprisonment in the tall, rough column of stone. Above thick lips and broad noses, the delicately narrowed eyes appeared infinitely serene. Gazing calmly over this violently raped country, they seemed to offer hope that a gentle divinity had remained to keep watch on the people.

Not too far from Angkor Thom was Angkor Wat, a working monastery where the people can still come to worship this divinity. A few of the successors of the Buddhist monks who long ago rescued this Hindu temple from the jungle still live here despite the efforts of the Khmer Rouge to obliterate them.

My heart sank when I saw how far away the graceful spires of the temple were from the entrance to the monument. They could only be reached by rolling down a wide avenue of large, irregular, stone slabs that stretched for a distance longer than two and a half football fields.

Deceptively smooth in appearance, this road proved to be a far more difficult surface than anything my wheelchair had previously encountered. Without care, the uneven edges where the slabs met could either jam my front wheels or allow them to drop suddenly. Knowing that either event could lead to my sudden ejection, I held on tightly to my metal armrests while T.C. and the guide took turns pushing me toward the temple. Halfway to our destination the sweating men offered me a chance to turn back but, having been informed that the temple dancers were to perform that day, I wanted to go all the way.

After what seemed a very long time, we reached the large stone-railed terrace in front of the temple where the performance was to be held. In an attempt to escape the scorching sun, I threw my wheelchair cushion down in a patch of shade by the terrace. Near me, the musicians were tuning up their instruments, and quiet notes like the soft calls of wild birds flew gently through the air.

Although I recognized very few of the instruments, the ancient sounds seemed to have a strange, primal familiarity, recognized in the dim caverns of my unconscious. Suddenly, a hush fell over the audience as random notes merged in approximate unity. Even a group of rowdy, jostling boys on a nearby stone slab began to whisper and turn their faces to the stage.

A young dancer in a long white dress, gathered with glittering red and gold at waist and neck, floated gracefully down the steps in front of the temple. As she bent to place a small cloth in the middle of the terrace, the tall, golden spires on her elaborate crown barely moved as she gently inclined her head. With ritual aloofness, she kept her elaborately made-up eyes either fixed to the side or straight ahead, making no contact with those of her audience.

Soon a slender woman elaborately costumed in fuchsia, turquoise, and gold slowly descended the steps to dance on the small square of material. On her carefully positioned head, a tall, slender spire arose like a unicorn's horn from an elaborate golden helmet. As she danced in solitary splendor, every part of her body subtly flowed from one carefully arranged pose to another, each graceful motion choreographed almost before the invention of time. Eyes, head, arms, legs, fingers, and toes all had ordained positions, some seeming almost impossible to attain with mere human limbs and digits.

Although her dancing was that of a young woman, her elaborate make-up could not hide the deep lines of age. Yet, even though the intense afternoon sun produced heat more brutal than that of the Amazon, her costume did not become stained with sweat, nor did her heavy make-up run in the heat. When a slender, extravagantly masked male dancer joined her, she rotated effortlessly on one delicately placed leg while holding the other up like a ballerina on a music box. As she held this position for many, long minutes in a spectacular feat of endurance, she seemed a creature of unearthly perfection, far beyond the discomforts of mere mortals.

With sweat pouring into my eyes, I watched this dancer with awe. Having survived the wanton destruction of 90 percent of her kind, she was the conduit through which ancient traditions had been safely preserved. When she left the stage, her place was taken by those who would carry them into the future: a half-dozen or so little girls.

Mimicking the gestures of their teacher, the little girls danced gracefully with fans, then bowls of flower petals. Dressed brightly in a multitude of gold-washed colors—fuchsia, red, orange, green, and chartreuse—they appeared to be about five or six years old but, in this country where food was short, they might easily have been three or four years older. As they concentrated on their dancing and carefully avoided eye-contact with the audience, they looked very serious and very young.

While the little girls danced, I heard a stir from the group of young boys at my side. One of their number had run around behind the temple to the opposite side of the terrace and was pulling himself up on the stone railing. Trying to get the attention of one particular little girl, he called to her and then ran through his repertoire of funny faces. Eyes carefully averted, she managed to maintain her distant facial expression until one of his comments drew a quick, wide-eyed glance. Beside himself, the boy immediately fell off the railing while his friends hooted and laughed. Given their boisterous normalcy, it appeared that, if adults didn't fail them, there might still be hope for Cambodia's future—one where little girls could dance and small boys could tease them in peace.

As we made our way back to the entrance, several Cambodian teen-agers crowded around to ask questions and seemed eager to learn about the mechanics of my wheelchair. Sensing opportunity, our guide began to teach one especially persistent young man how to maneuver it over the uneven, stone avenue. Suddenly, one front wheel dropped, nearly catapulting me from my seat. A loud, grating sound announced that there was a problem. Looking down, I realized the situation was serious. The hard rubber tire on my left front wheel had completely flipped off the rim. My heart sank as I spotted it a few feet away. Still a football field's length away from the entrance, the wheelchair was going nowhere, and, without it, neither was I.

As a crowd started to gather, T.C. quickly appeared at my side. As a growing crowd of men and boys discussed the situation, their expressions did not encourage optimism. Finally, however, with one mighty wrench of his hand, T.C. stretched the tire enough to flip it back on its rim. As I thanked him, I vowed to never again lose track of my front wheels, no matter who was moving the chair.

* * *

In the peaceful environment of Angkor Wat, the past troubles of Cambodia had begun to seem relatively remote. But on return to Phnom Penh, the screams of torture victims seemed to echo through the years and rise to meet our landing plane. When our guide met us at the airport, we asked if we could get seats on the next plane out.

"There's one leaving for Bangkok on the runway now," said the guide, not questioning our change of plans. "It should take off in five minutes, but I can tell them to wait."

She disappeared for a few minutes, then came rushing back with a uniformed man who checked our passports. He then escorted T.C. to a nearby desk as another man appeared to push me hurriedly toward the boarding gate. Traveler's checks were flying through the air as I signed them and tossed them toward the guide who trotted rapidly at my side. Suddenly, someone shoved my passport into my hand, the guide yelled good-bye, and I was outside. T.C. was nowhere in sight.

Alarmed, I said, "Where is my friend? We must wait for my friend." Ignoring my concern about T.C.'s whereabouts, a white-uniformed man lifted me into the front of a pick-up truck, then tossed my wheelchair into the back. Gears grinding, the truck rapidly accelerated and sped toward the waiting plane. As I looked worriedly back toward the terminal, I spotted T.C. loping on the tarmac several hundred yards behind.

Props whirling and positioned for take-off, the plane had already started to move down the runway but rocked to a stop as we approached. A door opened and two men ran down the small set of steps that had dropped out, and lifted me, wheelchair and all, into the plane.

After I buckled up, I started to worry because I couldn't see T.C. Becoming separated from him in a foreign country had been one of my greatest fears. I needn't have wasted the emotion. He had stayed at the front of the plane to make sure my wheelchair was securely strapped in place, and soon dropped smiling into his seat. Before he could get his seatbelt fastened, the plane took off. I felt as if we had caught the last helicopter out of Saigon just before the end of the war. Perhaps we had. In less than a month after our visit in 1993, United Nations forces, which had provided a transitional authority before the elections, pulled out. By the new year, the Khmer Rouge savagery began again and the narrow window of opportunity for Western visitors to see this country suddenly closed.

Once safely out of Cambodia, however, it didn't seem right to have left so suddenly only to avoid things I preferred not to see or hear. Even if I played ostrich, Cambodia's holocaust and ensuing problems still existed. Such running from reality seemed a very selfish luxury, when I thought of those like the beggar-woman who couldn't get out. No matter how I tried to justify my initial reactions, these people still deserved to be seen and heard.

Perhaps the world tries to look away from holocausts like the ones in Germany, Cambodia, Bosnia, and Somalia for the same reasons I had looked away from the beggar-woman and run away from Phnom Penh. No one likes to confront reminders that life is neither safe nor sane. It may often seem easier to run from people and things that threaten our idea of how life should be than to overcome uneasiness and approach them. However, if we choose to escape we not only lose the rich experience they have to offer but also risk loss of our own humanity—an outcome neither particularly sane nor safe.

Perhaps we should always look into the eyes of those whose condition makes us uncomfortable—the beggars, the ill, the disabled, or the downtrodden. If we don't have the courage to learn the lessons they have to teach, we can, of course, turn away. But if we make it a habit, there may ultimately come a day when we no longer have a choice about avoiding the beggar's eyes. Each morning we will find them staring out at us from our own mirror.

* * *

Cambodia was a reminder of how unsafe life can be or of how we run from reminders of this reality. It took a visit to neighboring Vietnam to discover how people can find a way to create for themselves the necessary bubble of relative safety, even as they live with life's harsher realities.

9

Dancing to a Different Drummer in Vietnam

The military-style pith helmets and conical straw hats worn by the crowds in the streets made them seem vaguely threatening when we arrived in Hanoi, even though, in 1993, our war with Vietnam was long over. Once these people had been "the enemy," easily recognizable in films and photos because of their headgear. If I had come here at that time, I would have been branded a traitor. Even now, eighteen years later, if I spent over one hundred dollars, I could still have been found guilty of "trading with the enemy."

As I watched the delicately featured people bustle along the gracious avenues and broad sidewalks of Hanoi, a city almost untouched by the war, it seemed hard to believe that we had battled with this country, nearly destroying it and our own social civility in the process. Although the war had divided our country like a prophet parting the sea, the Vietnamese seemed, surprisingly, to bear no grudge against us. When they spotted T.C. and me in our hired van, many smiled broadly and gave us a friendly wave.

The people of Vietnam seemed to be so busy hurrying toward their own delayed destiny that they had little time to worry about a couple of American tourists who, like them, had had little say when their leaders declared war. Their battle with our country had been one fight among many over the past thousand years. Over centuries, they had battled both against being led by neighbors and against colonizers from farther away, and insisted on marching to the beat of their own drummer rather than that of another.

One can select which drummer to follow either because one choice gives more pleasure or is not as bad as the alternative. Sometimes it's a toss-up. Which of these characterizes one's decision is a matter of personal definition.

I didn't know at first that such a decision was in my future when two men joined T.C. and me at our hotel dining table. As the slowly rotating ceiling fan squeaked a weak battle cry against the muggy tropical heat, we all compared travel routes through Vietnam. On discovering everyone's plans were identical, my heart sank.

Somehow, T.C. and I had been placed with a small group tour, instead of the private one we had booked. As a *fait accompli* in this very tightly controlled country, there was no way to alter arrangements already made for four visitors presumed physically able, or to rearrange the itinerary to accommodate my special needs for rest. At first, this did not appear to be a major problem, especially after an easy morning visiting Ho Chi Minh.

Like Lenin, Ho was of course dead, and like him he was bizarrely ensconced in a glass sarcophagus. Despite his wish to be cremated when he died in 1969, materials had been gathered from all over Vietnam until workers began construction four years later on his mausoleum. Brooding grayly over the huge, cement expanse of Ba Dinh Square, the flat-topped square memorial looked oddly like a negative image of the Lincoln Memorial, minus the statue of Lincoln.

As four Vietnamese guards bent to pick up my wheelchair at the entrance, then carried me around the well-lit bier, their slow, quiet steps were perfectly measured. The ride was even smoother than that given by their counterparts in Russia when I had circled Lenin's still body. As I peered at this quiet man carefully posed on his bier, I expected to see an angry, evil face—the face of an arch-enemy of my country. Instead, I saw only a fragile old man with a calm, wrinkled face, complete with a wispy white beard. Gentle in death, he looked so harmlessly ancient that it seemed more appropriate to use his original name, Nguyen Tat Thanh, than his more famous revolutionary alias of Ho Chi Minh.

Despite echoes of "Hey, Ho, We Won't Go," which still rang in memory from my student years, Ho now was the one who was gone. There was no doubt about it. Despite yearly trips to Russia for maintenance, his pasty-white face looked as if it were crumbling at the edges. It had shed more and more tiny flakes of skin with each passing year. Even the wrinkled hands calmly crossed on his chest could not totally divert one from thinking about the possibility of slow putrescence inside. Unlike Lenin's body, the body of Ho looked very, very real and very, very dead.

Spaced five paces apart around the inside perimeter of the mausoleum, the guards wearing broad Sam Brown belts and large, holstered guns were much more noticeable than those in Lenin's tomb. Perhaps they intended to be more obvious to make sure people followed the rules, which barred anything indicating disrespect or threatening danger. There was to be no use of cameras, wearing of hats, or placement of hands in pockets when admitted to this almost holy sepulcher.

All over Vietnam, security was to seem more overt than in other Communist countries I had visited. Sometimes I even wondered if our guide was a spy. Almost too interested in American culture and our home towns, his answers to questions about his culture were along pure Party lines, and he gave very little information about himself.

On our first day of touring, our small white van eased its way carefully through the heavy flow of traffic out of Hanoi. Multiple gaggles of bicycle riders threaded their way amidst the cars while motorcycles darted in and out like colorful stinging bees. Their flashing colors came not from the khaki clothing of the young male drivers, but from the bright pastel tunics of the tiny young women draped side-saddle behind them.

Contributing to the traffic nightmare, vendors frequently ran from carefully spread-out arrays of bright fruits and vegetables to dash into the midst of all the moving vehicles and make a quick sale. At the edges of the streets, women with faces nearly hidden by large, conical straw hats carried full baskets of yet more vegetables. As they carefully balanced them on slender yokes across their shoulders, they scarcely seemed to notice the danger to them from the impatient trucks and cars.

Cars had to swerve to avoid hitting one old man who walked very slowly at the edge of the pavement. Almost bent double by the weight of two large, latticed wicker baskets balanced on a sturdy wooden yoke, he seemed to struggle with his burden. As we passed near him, it could be seen why. One of the openings in his carefully woven carriers was completely occluded by the pink, quivering snout of a pig.

Gradually, pedestrians became fewer and traffic thinned as we neared the outskirts of the city. As paved roads changed to dirt, tumbling-down wooden shacks began to replace the graciously proportioned, neo-classical buildings left by the French. Ultimately, even the crowded-together shanties thinned out and gave way to flat, boggy rice paddies as we headed for Halong Bay. Maximally cultivated, the land's green evenness was broken only occasionally by the angular contours of tiny villages.

Suddenly, our rear tire blew out with a bang and the van jarred to a grinding halt. After unearthing a set of clanking, rusty tools from under the floor, our driver began a struggling exchange of a threadbare spare for the damaged tire. With a serious expression, the guide watched him for a few minutes, then told us we would have to return to a village we had passed a few kilometers back. There we would wait while they patched up the hole in the tire and, if they had the facilities and the will, have the bent rim repaired as well.

After our van pulled in near the single pump of the only gas station in this area, the men, including T.C., wandered off to explore. The guide stayed behind to push me to a small cafe where I could get out of the steadily increasing heat and buy some water.

The sun was so bright and the room so dim that approaching the cafe's square opening was like coming near a cave. As we entered, several unsmiling faces seemed to loom disembodied out of the gloom. In the sudden silence, the men froze in place like a tableau. Under chiseled scowling brows, their unblinking narrowed eyes stared in quick, icy appraisal. Uncertain of how hostile they might still be to Americans, I gave them a tentative smile. It was not returned.

As I grew more and more nervous, the guide barked a sharp word. Anxiously pushing back his thinning black hair, a man came hurrying out from his place behind a large, tarnished, silver cash register. After a brief conversation with the guide during which he frequently bobbed his head, he bowed his welcome toward me. Then, one long step across the cracked linoleum floor brought him to a quickly vacated table at the entrance. It appeared to be first in a line of four small vinyl-covered tables, but there may have been more further back in the murky depths of the small room. The ones I could see were crowded with men who watched intently as he removed one of the mismatched bentwood chairs to make room for my wheelchair. It did not seem a good time to express my preference for sitting in the chair, now removed from the table, rather than remaining in my wheelchair.

The atmosphere became slightly less hostile, but it was difficult to read the emotion of the men as I maneuvered to my place. This far out in the Vietnamese countryside where foreigners are rarely seen, it was quite likely these villagers might view me with suspicion. I was very aware that I was surrounded by men who had undoubtedly fought in the war and may have lost family to it as well.

Uneasy, I turned to the guide hoping to find something I could order to be polite, then tactfully make my escape. He said a few words to the cafe owner who was once again behind the register, and soon the man came back, nervously smiling. With a great flourish he placed a five-inch square packet wrapped in banana leaves on the table in front of me.

After glancing at the guide who smilingly told me that this was a Vietnamese favorite, I slowly began to unwrap the warm package, praying it would be somewhat safe to eat. Fortunately, inside was a semi-soft rice square which had been thoroughly steamed inside its leafy covering. Conscious of being watched by a multitude of eyes, I carefully took a small bite. Sweet and flavored with coconut milk, it was so good I smiled and nodded, then took a bigger bite. The quick relaxation of the observers was an almost palpable sensation as conversations softly resumed.

Upon leaving the restaurant, the guide went to search for T.C. and the others after carefully placing me in the shade. I was immediately surrounded by children. To my surprise, when they discovered I was from America, they quickly ran away. Only one toddler, eyes big and serious as she sucked her thumb, remained to watch. Mildly upset by their flight, I wondered what Vietnamese children had been taught about Americans. But within a minute they were all back, giggling and clutching books.

After they positioned themselves in an order known only to them, one boy opened his book and began to read. As he carefully pronounced the sentences in his English primer—written at a level slightly above Dick and Jane—he hesitated after each one to check my reaction. Slowly he read a written conversation in which Mrs. Jones, a teacher, was introduced to Dr. Smith, and vice versa. Whenever he stumbled over a word, the boy quickly looked to me for a correction, then carefully repeated my response. Soon a little girl, jostling him and competing for my attention, began reading alternate lines in the same text.

As I watched my delighted students, it was hard to believe that our countries had been at war, and pretty children like these had been strapped with explosives and used to destroy our soldiers. It was equally hard to believe that such innocent ones had ben killed by our bombs or, more gruesomely, by our fighting men.

I had been a student when a movie star was widely hated for coming to Vietnam and, worse, for liking the people—an action considered by many to be traitorous. And now, I was sitting in a small village outside Hanoi teaching English to Vietnamese children. I could find no way to sort out what I felt, but the overall feeling was good. All too soon, the tire and rim were repaired and we continued to Halong Bay.

Blue and pretty below a half-ring of hills, Halong Bay seemed like a magical lagoon in a fairy tale. As our boat's prow, elaborately carved like a dragon's head, pierced the peaceful waters, the experience was almost as mystical as being carried on a real dragon's back. Moving with dream-like slowness, we passed small junks and sampans appearing to float in mid-air where the blue of the bay and the sky blended as one. Surreal, through the light mist, tall rounded islets rose above the water like a collection of dragon's eggs in a flooded nest.

By legend the strange moonscape-like surface of this bay had come into being when a great dragon ran from his home in the mountains to the sea. As the water splashed up to flood the gouges made by his giant slashing tail, these odd humps of vegetation-covered chalk were all that remained above the water.

Halong Bay had received its name from the words *Ha Long,* which means "where the dragon descends into the sea." Among the patches of mist, it seemed almost as if I had traveled through time and come to a magic place where dragons still might live.

When we docked on a small island that jutted high above the boat, the fantasy was soon dispelled as men shouted at each other and tossed ropes. Within a few minutes our guide took the men on a climb up a long ladder of steps that zig-zagged up the steep slope. I had no choice but to remain behind.

If I had not heard somewhere else about the caves on top of these islands—about the stalagmites, stalactites, rock paintings, and buddhas that one might find—perhaps I would have not felt a great pang of loss upon finding that I couldn't reach them. But to be left behind by others who could see them, when I couldn't, was more difficult than I would have anticipated.

Soon, however, as I relaxed on the deck of the boat, I realized that the others were also missing out on an experience. The short time allowed for this excursion left no time for them to relax like me and soak up the atmosphere of the infamous Gulf of Tonkin.

My companions had not had time to study the duck's head, complete with white head and yellow beak, on the prow of the *Hai Au,* which had docked next to us. They had not had time, like I, to track the course of the small boats—sampans, junks, and more—with billowing sails that looked like halves of brown thumbnails split vertically down the middle. Nor did they have the opportunity for a close examination of the living quarters on the long, blunt-prowed canoes, which women poled back and forth near the dock. As we had motored by in the dragon boat, it had not been nearly as easy to peer under the half-cylinder-shaped shelters laying over the midsections of these busy boats.

As long as I didn't think about what I was missing at the top of all those steps, I enjoyed what I was viewing at the bottom.

But when they returned, the men all did a most unusual thing. One by one they all solemnly assured me that the climb hadn't been worth it and that I hadn't missed anything. Despite their apparent sincerity, their assurances rang false. But, glad to have them back, I quickly suppressed the question about why those in control of tourism had brought us here if there was really nothing to see.

Between Halong Bay and Hanoi was a car ferry that allowed people to cross the river despite the lack of replacement of blown-up bridges. No one except me was allowed to stay in their vehicle. Although I missed out on standing on deck like the others for a close view of the transit, I was again favored with a unique experience. Attracted by the oddity of someone apparently breaking the rules, six small girls ran up and arranged themselves in a half-circle on my side of the van.

During several attempts at conversation, it seemed that the only English word they were confident in admitting they understood was "American." But soon we found our common ground when the oldest girl asked me if I knew the French song "Frere Jacques." As we harmonized on the rounds, first shakily, then more strongly as we got the hang of it, I thought how beautiful it was but, yet, how odd. Only eighteen years after a war in which my country fought theirs, I was happily singing a French song with Vietnamese schoolchildren on a ferry made necessary because my countrymen had blown out their bridges.

When adults came to claim their parked trucks and cars the girls, waving happily, ran away chattering. By the time T.C. and the men got back, the girls were not in sight. Knowing no way to share the experience I said the only thing I could think of:

"You didn't miss anything back here."

From that day on, it seemed that there were two different tours running: the one I was on and the one everybody else was taking. The real separation began when we left Hanoi for

Danang. The pace of the tour was a quick-step rhythm of traveling one day, touring the next, then repeating the sequence. A rushed itinerary of "If this is Wednesday this must be China Beach" was not a comfortable one for me. Almost constantly exhausted, I could barely keep up.

All too quickly, the cities fell behind us as we moved down the Vietnam coast. The list read like a litany of famous battles: Danang, Hue, and finally Saigon. They all went by in a blur of fatigue. The things I remembered were mostly the beauty of the drives—from city to city or from the airports to our hotels. Although the others saw temples, trod beaches, climbed through tunnels, and bargained in the markets, I mostly saw only the insides of our lodgings.

On our final night, I requested an adventure that probably wouldn't have won by a majority vote. Possibly feeling sorry for my having missed so much of the tour, our guide and the group agreed to have our final dinner at a most unusual restaurant when I mentioned I'd like to try some really special food.

The first sign of anything unusual about the outdoor restaurant was what seemed to be a small animal zoo just inside its perimeter. Upon seeing the menu, I realized that this was no zoo. The denizens of the cages were dinner, to be selected like one picks out one's lobster in a seafood restaurant before it is boiled and served in butter-dripping glory.

As our two travel companions desperately searched for steak or salad, or at least something familiar, T.C. and I cheerfully ordered bat and cobra. The experience of picking out a fat bat from the small cage where they hung, was exceeded only by the process of approving our cobra.

A hush fell over the restaurant as a slender man walked over to our table holding a cobra tightly by its neck. As I tensed and reclined my chair as far away as I could, he smiled and carried it to a small area clear of diners where he quickly threw it on the floor then jumped away. Weaving his body about to get its attention, he danced near it, then far. Suddenly it coiled and struck.

Staying at the bare minimum distance for safety, the man laughed, then goaded the cobra again. Hissing, it flared its hood then aimed for its tormenter like a hurled javelin only to fall short of him by inches. Again and again, it struck and missed. Once it tried to slither away, but the man wouldn't let it, kicking out and dancing to the front of it. Finally, after one strike too close for comfort, his hand moved toward its head with a blur of speed. A collective sigh arose as he grabbed it just behind the vicious fangs.

Holding the deadly snake up at arm's length, the man nodded to a waiter who came to me and asked if I approved. Still stunned by his performance, I nodded, then finally remembered to turn and smile at the man. Looking relieved, he marched triumphantly off to the kitchen where the boiling pot waited. Those at the other tables burst into applause.

I had to admit the tough cobra, served in small segments, was good, but not as delicious as the plump bat. As T.C. and I chewed our meal, savoring every bite of this alien food, our travel companions could only look away. When they returned home they would not share my memories of this dinner.

As I relished my helping of bat, I suddenly realized that we had all been on our own trips even when physically in the same place. I had enjoyed my own experiences and they had enjoyed theirs. Even if I had been able to join the tours I had missed, the situation would have been the same. By winding up on the wrong kind of tour in Vietnam, I had learned, as never before, that we all must travel through life on our own individual journey and these journeys are never the same.

When you hear the beat of someone else's drum, you may, of course, attempt to march according to its rhythm. But that hadn't been good enough for the Vietnamese throughout their long history. And on this journey to Vietnam, it was not all that good for me. Perhaps after all was said and done, I *was* guilty of "trading with the enemy" because I brought home a secret they may have known for centuries. In their beautiful tropical land, I had discovered that once you disregard the drums of others, happiness comes when you dance along to your own.

*   *   *

Sometimes the drumbeat urges you to dance straight ahead and sometimes it impels you back up to try another step. On a second journey to India, I explored several alternate routines as I learned the many ways I could dance to my own drummer.

10

To Control One's Destiny: India Revisited

I never thought of dancing to my own drummer as dangerous. But that was until I was driven north from Delhi to Corbett Park where I hoped to hunt for a tiger. While waiting for T.C. at a rest stop, I was suddenly confronted by totally different beings with an even more menacing disposition. As a small group of villagers began to gather about me, an ugly destiny seemed to leap toward me from behind their eyes. It snatched any illusion I might have held of having control of my fate as surely as a tiger grabs its prey.

Many of us cherish the illusion that we can control our own destiny—feel that we must to be happy and safe in the world. To maintain this deceit, we often dance away from an unpleasant circumstance and hope it will just go away. Sometimes, unable to get away, we are forced to relax our grip on the reins of control and hope circumstance will take us where we want to be. And sometimes, we can get beyond the need for control, and come into the true greatness of the human inheritance. I was to do all of these on this trip to India. But on that hot, dusty road that first day, all I could think to do was to run away from what was suggested by those angry eyes.

We were in traditionally Muslim land—male-dominated, secretive, and hostile toward women. On my request, our driver had stopped near a small rickety outhouse a few feet from the road at the outskirts of a desolate village. The area appeared to be deserted. Afraid my wheelchair would become mired in the sandy-soft dirt, I had used a cane to hike alone to the ramshackle facilities, while the men wandered down the road to a small stand of trees.

First to return to the car, I leaned against the warm metal, daydreaming of tigers and enjoying the chance to be vertical. Suddenly I became aware of the small, muttering group of young men who stood staring at me from a few yards away. As they surveyed my T-shirt-and-jeans-clad body with increasingly hostile eyes, they became an increasingly ominous presence.

Hoping to ease the rising tension, I greeted them in Arabic: "*Salaam Aleikum.*" It was the only phrase I knew. But, already chilled by anger, their eyes became even more coldly venomous. Their hatred seemed almost a living thing as it jumped the distance between us.

My mind screamed the realization that I had seen very few women for the last couple of hours. There had been no females strolling the streets or standing casually in the doorways of the small villages we had come through. Two women riding on the back of an ox-pulled cart, the local form of public transportation, had turned quickly away from our passing car. Huddled together, they had been shrouded, from the crowns of their hidden heads to below their invisible knees, by thick, black veils. It was hard to imagine how they could see through such a covering, so opaque that barely a hint of human form was revealed within.

With a bare head and in T-shirt and jeans, I was dressed modestly enough by Western standards. But it was hard to imagine how I appeared to the staring men who lived in this world where women were so totally hidden from public view. The look in their eyes was one I could easily imagine in the eyes of a mob ready to hurl stones with deadly force—the scripturally prescribed punishment for a loose woman.

Finally hurrying toward me past the growing crowd, the driver seemed nervous as he asked me to get in the car and lock the doors. I quickly complied. As the driver went off in search of T.C., the grim-faced young watchers stood very still. When T.C. returned with him, smiling, their muttering grew louder. Wondering how anyone could be safe in the desolate countryside, I still felt only partially secure after we sped away.

Only a few miles up the road was another small, dismal village where a different type of angry crowd had gathered. Shrieking and hissing, a group of desperate vultures fought frantically over something hidden amidst their scrabbling feet.

Perhaps they were making a meal of one of the emaciated, predatory goats that moved spastically about the village, anxiously searching for something to blunt the sharp edges of hunger. Perhaps it was one of the skeletal yellow dogs skulking about with side-long glances and lowered heads. No gentle pets, they looked only a morsel of garbage away from starvation in a village too poor for garbage. Remembering the dead man on the muddy hill in Agra, I wondered if the meal was something even worse. I had no desire to find out.

Even when we arrived in the park, it was difficult to shake a general impression of desperation and desolation in this part of northern India. Our accommodations, much more spartan than expected from the guidebook description, closely resembled a deserted concrete bunker, and the inside was even worse.

The thin, foam mattress on a hard narrow cot was covered with a sheet, grey with dirt and so threadbare it was shiny. Green mold and rust competed for every wet space in the bathroom, while badly torn mosquito netting over unglassed windows offered no illusion of protection. Only the thought of the long drive back through now-darkened desolation kept me from heading straight back to Delhi.

I had never seen a place so obviously promising disease.

Even boiled coffee was not safe to drink, so of course we could not eat the food served at this primitive lodge. But, having packed a water filter, insect repellent, and granola bars—always useful parts of a tropical kit—we stayed and made the best of it. I was still determined to have my adventure.

After a surprisingly good night's sleep, the future looked much brighter. The only problem, as I saw it, was that I had no idea as to how I was going to get on top of the elephant I was to ride while hunting for the tiger. Having managed to come all this way, however, I decided to just trust circumstance and see what would happen.

In reality, getting on the elephant proved to be very easy. Moving slowly in the chilly dawn, three ponderous pachyderms plodded slowly into the main courtyard, then waited patiently for their riders next to an elephant-high flight of wooden steps. When my turn came, all I had to do was to let T.C. help me climb up to the open platform at the top, then just sit down on the elephant's back.

Grateful that my part had been so simple, I quickly grabbed the corner post on the square, canvas saddle when, on a signal from the mahout on its neck, our elephant slowly set out for open savannah. I held onto the polished wooden column very tightly. Now that I was on, I didn't want to disgrace myself by falling off.

Swaying from side to side like a ship amidst high waves, the massive beast carried T.C. and me farther and farther from our small island of civilization. As we moved slowly through the dense, high savanna, unmarred by human presence, we entered a world of rare beauty, hot and dreamlike under the intense rising sun.

Riding in these lush grasslands, I was no longer a disabled tourist. In my heart I was a powerful maharajah setting out with my servants to brave a snarling tiger. My dirty, white saddle became a gilded howdah, sparkling with gems and filled with soft cushions, and my old Canon camera was now a sharp-pointed spear.

The only thing missing was the tiger. Even in this park famous as their home, finding one was not going to be easy. As they had been hunted nearly to extinction, there were fewer than a hundred left in this part of India. I'd be lucky just to catch a glimpse of one.

Although I never could comprehend why people would want to kill such a magnificent beast, I soon understood the thrill of the hunt. Searching intently for our prey, I could feel the excitement a maharajah must have felt as his "beaters" whipped the grass to flush a tiger into the open. Listening carefully, I fancied I could still hear their excited shouts wafting faintly over on the breeze.

Suddenly signaling for silence, the mahout tapped my arm then pointed to a fresh pawprint. It was as big as a dinner platter. A very large tiger had recently passed through the tall grass and might still be near. It might even be hiding, silent and hungry, to watch our every move.

Not daring to make a sound, I froze in place as the mahout slid down the elephant's trunk and knelt to scrutinize the spoor. After an intent inspection, he signaled the elephant to wrap its trunk around him and lift him back up to his seat between its ears. He pointed ahead to where a deep ravine, choked with rocks by the rage of a long-ago river, cut abruptly through the grassy terrain.

With beginning disappointment over what I thought was the loss of the trail, I scanned the rest of the surrounding savannah without much hope. Suddenly I clutched my saddle-post with a death-grip as the mahout, not giving up so easily, urged the elephant straight down the steep, stony incline of the chasm.

As the massive animal carefully picked its way across rocks the size of small boulders, small pebbles loosened by its heavily cautious steps rattled in abrupt, startling showers to an invisible, unheard bottom. The elephant's strong, sturdy legs began to feel almost as unsteady and erratic as my own as it carefully tested each of the jammed-together rocks with a ponderous foot, then slowly shifted its tremendous weight.

I had once heard that an elephant will not step on anything that won't support it. Clutching this knowledge as a straw of hope, I tried to relax as my elephant slowly climbed down the slope of the dried-up riverbed, then up the other side.

It was an exercise in complete trust, which I suppose comes to all people at one time or another, perhaps a little more to those in wheelchairs. I was in a situation over which I had absolutely no control, nor could I readily acquire it. But I needn't have worried. Carefully placing its feet, the elephant didn't stumble and, clutching my post for dear life, I didn't fall. Together we climbed out the other side of the ravine to emerge triumphant on level grassland.

The mahout inspected the ground, then nodded happily. We were apparently still hot on the trail. A little ways ahead, the high grasses rustled briefly, then as quickly became still. The elephant seemed unusually reluctant to proceed. As the mahout gave it an extra nudge, and then one more, I wondered if the tiger was hiding in the light-dappled shadows.

We went a little farther, then the mahout shook his head. The trail, always faint, had completely disappeared. As we turned to go another way, I carefully scanned the tall grasses one last time. Then I saw it. Off to the side there was a quick flash of yellow. And then it was gone.

I'll swear to this day that it was the tiger.

Under the now hot Indian sun, the giant cat remained elusive. Although we hunted for another hour, we never caught up with it again. But I had captured a trophy even more rare and precious. The sight of that stealthy movement, the ride on the elephant, the excitement of the hunt had become images emblazoned forever on my mind. A far better trophy than the snarling heads or tattered pelts once brought back by the maharajas, they were priceless reminders that dreams can actually come to life. Their birthing had been made easier by my acceptance of circumstance instead of insisting on control of the details.

But the best was yet to come. It took the peace of a journey to Varanassi, the holiest place in all of India, to reveal that to seek control is to chase after a useless mirage, and has little relation to finding true happiness.

T.C. and I left our hotel before dawn to go to the banks of the Ganges. After our driver parked where the paved road ended, we were confronted with a few hundred feet of slippery, sloping earth, not quite dry from the heavy monsoon rains of the previous night. To get to the water, we could either traverse the wheelchair-unfriendly surface or attempt to negotiate the narrow, steep set of steps going down the hill. Neither option was particularly appealing. Although the sun had not yet risen, an enormous, milling throng of colorfully garbed people—gossiping, bartering, and worshiping—created a nearly impenetrable barrier on either route chosen.

After a debate, T.C. and our guide decided not to risk carrying me down the nearby set of steps. Instead, they began a slow, laborious descent, steadying my wheelchair over the uneven surface as they forged a path through the surging mob. Watching us with interest, the pilgrims we came near kept a careful distance away as they stared, or smiled acknowledgment of our presence.

Near the water's edge, we came upon a treacherous span of boggy ground where the wet tracks of those returning from the river mingled with the remaining dampness from the night. Just as my wheels began to sink, several helpers jumped forward with encouraging shouts to help T.C. and our guide boost my wheelchair high into the air. Grunting and sweating heavily, they struggled to carry me over the sucking mud and across a large cleft in the earth that loomed just ahead. I held on tightly and tried not to look at the hand of the man on my right, perilously near the lever that could suddenly jettison the leg-piece. Praying not to be dumped into the filthy mud, I gritted my teeth and didn't distract him with a comment.

Once we finally reached our boat, already pulled up a few feet onto the riverbank, the men lifted me, wheelchair and all, into it without ceremony. Wedging my wheels tightly between cross-pieces of the boat, the men left me there, stable and secure, as the boatman fought the heavy current of the flood-swollen waters to row us into the middle of the Ganges and head downstream.

Dirty with a sediment it was best not to look at too closely, the night-black river began to turn a lighter tan as the sky turned pink and gold and blue with the dawn. As the sun stole over the horizon, it all too briefly became a river of gold. A soothing mumble could be heard from the hundreds of people standing waist deep in the water near the river's bank as they splashed it over their heads with a rhythm that was sometimes ritual and sometimes bath. A gentle lowing came from the hump-backed cows who stood only a few yards away in the shallows.

As we moved down the river, the slap-slap of the day's laundry being beaten on the steps of private docks by brightly saried ladies came faintly over on the breeze. It provided a rhythmic complement to the melodic laughter of children who ran in and out of the bright-colored buildings lining the riverbank, or dived from the *ghats* and played in the water. Like a gentle drone accenting more complex harmonies, the river softly slurped at the high walls of their homes as well as hotels, and temples made splendid by bright paintings of mysterious gods and goddesses.

Occasionally, rough, brown umbrellas, like giant, fibrous

mushrooms, were clustered together on the open slope near a public *ghat*. Under their shelter, dozens of priests waited for those on the dock who might need their services. The Ganga, as the Indians call it, was not only a holy site where pilgrims came to worship, but the most sacred site where the dead can rest. All of these actions required a ritual, for which priestly guidance was needed.

Appearing strangely alone in a gap among the crowds, a body wrapped in red and gold silks lay with its feet in the river as we passed. Four sober, bearded men reached to splash the holy water over its head and chest, before carrying it gently away. Even this early in the morning, the faint, sweet smell of burning flesh from the village crematorium was already beginning to permeate the air. Ashes would be brought back to the river for the most holy of all burials.

It all seemed so right. Men and Women. Health and Infirmity. Life and Death. All represented on the holy river as the pilgrims came to it at dawn. All part of a continuity, which neither wanted nor needed my input to arrange itself in the way it was meant to be.

I wanted to submerge myself in the sacred water to ritualize my inclusion with all others in this cycle. However, when I thought of all the profane substances that daily found their way into the river, I settled for taking away a small vial of the water. Later I poured it over myself in my room's shower where it could be quickly sluiced away.

It had been a dream to come to Varanassi and fragments of not only the realized dream, but perhaps of a more ancient one, remained in my consciousness for a long time. They had been gently lured into staying in place, like a tethered will-o'-the-wisp, by the priest who placed a lotus blossom and a bag of herbs in my hands before leading me through the Sanskrit words of a healing ritual. The memories were fastened in place forever by the powerful magic of the dreamscape we crossed when we left the holy city.

The monsoon storms of the night had made way for a torturously humid day, when we took off into an uneasy sky choked with thunderclouds. As our small jet flew higher and higher into the damp-clogged heavens, its screaming engines seemed unable to pull it up from amidst the towering black cloud-anvils and into the sunlight.

The pilot banked left, then right, then left again as he skirted one deadly, black storm-cloud after another. Any of these treacherous heaps of vapors could easily ingest the plane, toss it about, then spit it disastrously from the sky. We were surrounded by a strange, ethereal beauty as the plane wound through the deadly canyons where lightening flickered a preternatural suggestion that we were not far from the Other Side.

Suddenly the plane broke free into a magical new world. An infinite, rolling plain of light-dappled billows stretched into the far distance where dense peaks of vapor rose like mountains into a crystalline blue sky.

The landscape of clouds seemed for a moment to be a place where one could dream the right answers to all the questions in life. But then the plane tilted downwards and descended into the bustle of the ordinary world. Sadly, the answers went back into hiding from mere mortals, exactly where they belonged.

Each early evening at my East Coast home, I'll suddenly stop what I'm doing and think, "It's dawn now in Varanassi. They're all at the river, and life is proceeding as it should." There's no need to control it, and no need to do anything more than let it happen. And as I join them at the holy river in spirit, I know there is no better way to be content than to relinquish the illusion that one needs to be in control—of one's body, one's world, or of anyone else. It's the way to let one's spirit rise free of it all.

* * *

A year later, however, a close-up look into the wild eyes of an African leopard brought another perspective on freedom, as well as a ripping, grand adventure.

11

Vulnerability and Power in Tanzania: Balancing an Important Equation

Cawing and shrieking near the edge of the dirt road, the rapacious crowd of vultures created an almost deafening din as they plucked at their prey. They seemed far more energetic than their starving counterparts in India. Stabbing their long, naked necks back and forth as rapidly as a snake flicks its tongue, they struck at their competitors almost as often as they tore at the meal in their midst.

Some leapt, fluttering wildly, onto the backs of those ahead of them, only to be driven back hissing, while the luckier ones hopped off with torn strands of intestine clasped tightly in their beaks. Crouching a safe distance away was a solitary hyena. Despite the tremendous power of her jaws, with no sisters to support her she dared not challenge this hunger-maddened coven.

Like any murderous mob, their numbers grew and grew. Circling down like crash-landing gliders, they came from afar for a share of the prey. As newcomers arrived, the many battle lines shifted to include them. As competitors realigned, the zebra-striped shape in their midst was briefly exposed.

Twitching with each grasped gulp of its flesh, the dead zebra was the impotent focus of this brutal assembly. I wondered how it had been singled out from its peers to fall victim to predators and those who scavenged their remains. Had it been too slow because of an injury or illness, or was there a hidden weakness apparent only to those who sought it? Perhaps in one fateful moment, it had simply been unable to blend with the rest of its peers in a confusion of stripes, and therefore been the one selected for this brutal end.

A chill came over me as I realized that all of these reasons for vulnerability could be applied as well to me. My senses heightened as surely as those of the gazelle who flicks and swivels her ears in the face of an unknown danger, then subsided like hers when no current threat was detected. Yet, frozen in fascination, perhaps like she would be by approaching headlights, I watched for many long minutes.

Barely past the border, I was thus introduced to Tanzania, where parched plains and fertile savannahs offered endless lessons in vulnerability and in power. Both blend savagely in the dramas played out by predator and prey, as well as merged within the individual players. In this home of the first humans, their descendants are offered a chance to make their peace with this heritage.

We had traveled only a few miles past the vultures when our driver again pulled to a watchful halt. This time he kept the motor running. About to cross our path was a big bull elephant. Spotting us, he came to a dead stop, and his ears began to beat a slow rhythm against his massive head. Burning with hostility, his eyes were like deep-set coals embedded in a pewter grate of wrinkles as the rest of his herd crossed the road ahead of us.

Moving slowly, the females traveled in small compact huddles, never veering far from their companions. Almost hidden amidst the dense forest of legs, several small baby elephants trotted rapidly to keep pace. As they came abreast of us, their mothers began a subvocal rumbling and the herd increased its pace. Watching us carefully, the bull responded with a swallowed growl, but remained in place.

As his ears increased their nervous motion and his trunk began to swish like a violent pendulum, I wondered what it was like to be the biggest and strongest creature on land. Yet even this birthright of indisputable power had not kept him from once being vulnerable when young and inexperienced, like the little ones the herd so carefully protected. Nor would it save him when he grew too old to protect himself. But right now, in his prime, he was the very essence of pure power.

The male seemed to grow ever more nervous when the last of a few stragglers finally made it past us. When we remained in place, now flanking the rear of the herd, he decided to make his move. Raising his trunk, he let out a loud, resounding screech, then began to walk directly toward our van.

After the first few slow steps, he shrieked again and began to pick up his pace. Alert for this possibility, the driver gunned the engine and outraced him for a few seconds. But as the bull elephant kept increasing his speed, then broke into a trot, I knew it would be close. Against the momentum of his charge, the metal shell of our vehicle would be little more protection than tissue paper.

Suddenly, he slowed back to a walk, again swinging his massive head back and forth. Then he subsided. As he came to a halt, I let out my pent-up breath. The danger was over. Motionless, he watched until we had driven out of sight.

The big bull had needed his power only to send us off, not to totally defeat us. Aware of our vulnerability, we had been only too happy to comply. In accordance with a natural preference of both parties to avoid injury, a balance had been struck. However, even among those of virtually equal power, even if that power is great, a balance is not always possible. Shortly after we escaped our confrontation with the elephant, to our surprise, we came across a dying hippo.

By size a powerful creature, the hippo has few enemies.

Placid by nature, it would seem that it would be rarely confronted by any adversary. Yet, lying in a shallow depression in the sand, once a life-giving pool but now holding barely an inch of water, the hippo didn't even have the energy to open its eyes. Only a slight motion of its sides, coupled with the absence of circling vultures, suggested it was still breathing.

The reason for its moribund state was brutally apparent. At the side of its chest was a large deep hole. At one edge of the gaping wound was a tiny trickle of blood which had already dried in the heat. It had been punched out by the bite of another hippo, our driver informed us. In a battle for the right to pass on its genes, this one had lost. Although its death was in accordance with the law of nature, I felt sad. After studying the quiet beast for a while, I wondered at its dignity as it conveyed a dying message that no one had power so great as to be completely invulnerable, even when it is at its peak. Then quietly I asked that we move on.

As we neared the Ngorongoro Crater, the flat, parched land changed to voluptuously verdant hills. I wished the hippo had, in its last moments, been able to receive the comfort of the quiet pools there, which were fed by a still viable source of water. But many other animals also needed this precious liquid and had found their way here to seek respite from the arid valley.

Well-camouflaged amidst the beige-brown tree-trunks as they reached for the highest leaves, giraffes were easy to spot only when they lazily turned their heads to view us. Browsing in the clearings were innumerable gazelles, eland, and topi, while dozens of stiff-tailed warthogs ran busily among them all. When I expressed surprise to our driver at finding so many different types of animals so peacefully grouped together, he noted that it was no problem for them, as long as they were all vegetarians.

After the obligatory laugh, the greater message made itself known. It reiterated one I had learned in the Amazon: one's vulnerability can be lessened by grouping together with those of similar intent. Unfortunately, however, in the human world, the best associations are not always as simple to identify.

After the long journey to the Ngorongoro Crater, I was amazed to find we were to enjoy five-star luxury at our hotel, whose rooms lined up like voyeurs at the edge of the crater. Even more amazing, the hotel was equipped with ramps throughout its three levels. Although a little too steep for me to manage on my own, these service passages proved no problem for T.C. to push me about.

As I looked out the window of my room shortly after our arrival the setting sun gilded the edge of the perfect geologic bowl before me, rosying a lower ledge before purpling the cavernous drop to the flat plains at the bottom. Feeling like a god, or an angel peering down from this most beautiful of all places, I strained to see the animals who had lived generation upon generation at the base of this dusty sphere. Hidden by distance and the growing darkness, their presence could only be sensed. Even though I couldn't see them directly, I was subliminally aware of their activities at the bottom of the crater: the big cats were waking for the hunt and small leaping creatures were making ready to flee their greedy teeth and claws.

It was only the next day, when we descended into the crater, that I realized that on safari I was as able as anyone else. No one—I mean *no one*—without experience and a rifle, *walks* down inside the Ngorongoro Crater where lions, leopards, and cheetahs stalk the thousands of zebras, wildebeests, and gazelles trapped there with them. As the smell of blood drifts across the flat plain at the crater's base, vultures scream and stake out the remains, while hyenas come galloping from miles away for a share. For obvious reasons, all visitors must be driven about in a Land Rover, specially designed for African roads.

Protected in a metal and glass shell complete with heavy-duty tires for a quick get-away and a lift-up top for good game viewing, I was no more vulnerable, nor more safe, than any other visitor. Like them, I went out on safari once or twice a day and could trade stories of the "kills" we had located with the best of them.

It seems an odd part of human nature that everyone, T.C. and myself included, was more eager to tell of what they had seen killed than of what they had seen wandering beautiful and free. Even knowing the end result, there was no way to avoid fascination with the lithe, quiet predators and skulking, eager scavengers.

On our first day in the crater, an unusual combination of savagery and peace drew our attention like honey draws bears. Near the dirt track that passed for a road lay six sleeping lionesses, lying on their backs like so many large, contented pussycats. Clotted blood streaked their distended stomachs and reddened their flaccid muzzles. Behind them, two males, snarling mutual warnings, kept careful distance from each other as they tore at something still and shapeless. A bit of beige fur still clung to the once-living thing as their raspy tongues hurried to scavenge any escaping drops of blood or crumbs of flesh.

Somehow it was not horrible. Instead, it seemed strangely appropriate that molecules that had only recently been part of a gazelle would soon re-form into claws, paws, or a mane. It was all part of an eternal cycle set long before my birth. But I was glad that, today, I was not the collection of atoms being rearranged into something else.

Suddenly another safari van arrived, and then another. Like ants streaming to a delicious morsel, van after van pulled in and jostled for the best place. Accompanied by the click of cameras and chattering voices, a huge cloud of exhaust fumes began to drift over the pride as the noisome vehicles completely encircled it.

The dining lions barely turned their heads, while the sleeping females raised theirs for a look, then dropped them again. Less vulnerable to natural dangers than most, the king of beasts had unfortunately been conditioned to disregard the one to which he was truly vulnerable. I wondered how long it would be before man's lack of concern and experience with maintaining a proper balance between power and vulnerability would destroy even this proudest of animals. As soon as we could find our way clear, we left the area. A few miles away, however, a pride of lions managed to shift the balance of power in their favor by assertion of simple, unadulterated dignity.

At first it appeared that our driver had slowed for no reason in the middle of an empty track through a desiccated area of the plain. Motioning for silence, he pointed into the far distance where there was a moving line of something almost impossible to make out. It was a very long line and it was slowly coming in our direction. Carefully eyeing its course, the driver then pulled up to where it should intersect the road, and cut off the motor. It was a few minutes before we could make out individual shapes in the slowly advancing procession. But soon we discovered it was a long string of lionesses and their cubs, as well as a few lions too young to have a mane. No large-maned male appeared to guard them.

"Where are they going?" I asked.

The driver shrugged his shoulders, spreading his hands helplessly at his sides. "They just go," he said, and said no more.

Slowly they came closer and closer. Just as it seemed they were headed right for our vehicle, another vehicle roared around us and pulled up just ahead. In a very empty country, the roads can suddenly become very crowded when even only one other car is near. However, it was in peaceful fascination, not grim competition, that we all silently waited for the lionesses and their young to approach.

Upon noticing the vehicles directly in their path the heretofore unwavering string began to break up. Some lions did not abandon the chosen course and walked directly up to the jeep ahead. Others veered off to the side and stood watching. A few walked over to the shade of a tree a few yards to the side of us, then lay down and closed their eyes.

As more and more lionesses reached the road, the pressure of the new arrivals finally seemed to give courage to their predecessors. First one, then another, and another walked around the front vehicle. Finally, either singly or in groups, the thirty-five lionesses and their entourage had all passed. In the distance we could see them reform their long straight line and watched until the moving tan band blended with the sandy color of the earth. I never found out what drew them.

Until now, we had been merely onlookers in the wilderness. Except for the confrontation with the elephant, we had never been in any personal danger, and the elephant had fortunately been relatively easy to escape from. Personal vulnerability had been compensated for by mechanical power and the knowledge of an expert scout. But the balance of power can shift abruptly, as it did when we traveled from Ngorongoro province to the Serengeti.

We were making good time on the dusty road until suddenly, as we tried to climb a small hill, the tires began to spin. As the wheels dug themselves deeper and deeper into a pit of drifted sand, our van quickly came to an exhausted halt. We were stuck as surely as a vehicle that has had the misfortune of coming to a dead halt at the bottom of an icy hill. Rocking the vehicle back and forth, the driver muttered under his breath as sand sprayed futilely behind us. Then he turned off the motor.

To my surprise, I discovered the van had no two-way radio with which to summon help. I tried and failed to remember meeting any other vehicles along the way. Whatever help we got would be gotten from ourselves unless we were blessed by fortune. Although common wisdom would suggest that in this situation I should have felt afraid, I instead enjoyed the sudden opportunity to be a part of this wild land.

As the wind whistled across the vast empty plain, the driver knelt to scoop handfuls of sand from the wheel well. I headed for the leeward side of the vehicle and sat down. There was nothing else I could do. Hugging my arms around my knees, I gazed at the peaceful terrain.

With a start, I realized something was staring back at me. Crowning the top of a small hump of earth about fifty feet away was the bleached skull of a wildebeest. Black and empty, its gaping sockets seemed to peer directly into my squinting eyes. I wondered briefly if I was in any danger of sharing its fate.

When a soft grunting sound came from what I had thought was a group of rounded brown rocks in the far distance, I turned my attention to the horizon. Moving closer in a long erratic string, the "rocks" revealed themselves as a large herd of wildebeest, flanked by a few sentinel zebra who acted as their eyes and ears. With poor eyesight and hearing, they, like me, sometimes needed more able helpers to survive. Despite the evidence of the skull that it didn't always work, they still remained placid in their faith. When I turned to check on the working men, I prayed to remain the same.

As our driver struggled futilely to keep the coarse beige granules from slipping from between his fingers, T.C. found a tiny shovel wedged in the trunk and joined in the effort. With a wind blowing ever stronger, the obstructing grains were replaced almost as fast as they were removed. The men worked faster and then began discussing what they could use to wedge under the wheels. The lowing herd of wildebeest drew closer and closer.

Stripping off his shirt, the driver jammed it under the left rear wheel, then ran to sit behind the steering wheel. Suddenly he revved the motor as T.C. gave a mighty shove from behind the van. The wheels gripped, but only for a second. Spraying sand, they renewed their spin before the driver could cut off the gas. T.C. quickly ran forward to wedge the driver's shirt more tightly beneath the stubborn wheel, then they repeated their maneuvers thrice more. Hope was fading fast. But all at once, with a shout and a shove, the van was free.

As we cautiously made our way down the road, we were soon forced to stop when the wildebeest reached us. Despite the hesitancy of their wiser zebra guards, the dull-eyed beasts plunged directly into our path. Instead of driving them away, a honk of the horn panicked them into running even more rapidly in front of our wheels. For ten minutes we waited as they all sorted themselves into single and double file and, after crossing the road, headed north.

These were the stragglers left behind by the more massive herds who had crossed the border a few weeks before. Following instincts built into their genes, they were leaving this land parched with autumn drought to seek water in still verdant Kenya. Vulnerable by virtue of being away from the main herd, and not too smart to boot, the instinct to stay together, honed for generations, seemed to be the only thing that kept the balance between vulnerability and power more or less in their favor.

One afternoon at the hotel that was our base in the Serengeti, I was too tired to go out on safari, but to my surprise this gave me an interesting advantage over other visitors. After the vans were out with the hotel residents, T.C. included, I quickly discovered that my room was not a place to linger. Shortly after lunch the power, produced by a massive gas-guzzling generator, was cut off to save expense. Not only was there no cooling fan nor light by which to read, there was no hot water to remove the dust of the morning road. Soon, I went out to recline on a shady veranda. In the valley below, a different side of life in the Serengeti played out before me as if on a theater stage.

High-pitched squeals drew my attention as, biting and kicking, a large male zebra took on another of his kind. Fighting furiously they occasionally disappeared in huge clouds of dust as they sought each others' most vulnerable parts. Wheeling and bucking wildly, they both managed to keep away from serious injury as they settled their dispute. Finally the loser ran off and the victorious male turned his attention to a watching harem of females. But when he approached the female of his choice, she trotted quickly away. Again and again he approached her. Again and again she refused. But finally, turning her back to him, she stopped and allowed him to mount.

While other visitors were out seeking "kills," I was seeing the other half of the life and death equation that was constantly being balanced on the African plain. But I was yet to learn that it is not only the power of reproduction and group alliances on the one side, versus claws and teeth and horns on the other, that helps to balance this equation whose solution often depends on the relationships between power and vulnerability. Two leopards were later to show me that having a strong purpose, coupled with the development of one's own unique nature to its fullest, may also be an important part of evening the score.

One day we traveled into the far reaches of the game reserve to one of its stranger places. Here, giant piles of boulders, scattered randomly about the flat, naked land, created a bizarre moonscape of sand and rock. Odd in their isolation, the boulders in some places looked as if a giant had tired of a game of marbles and mounded them into little heaps to wait for another day; in others as if the giant had tried to create a Stonehenge playhouse, then run out of obelisks.

All the small gatherings of wandering animals—leaping Thompson's gazelles, clots of kudu, and a few remaining wildebeests—seemed to avoid wandering too near these barren heaps of stones. Then I saw the leopard, and knew the reason why. Lying relaxed at the very top, she kept careful watch over all the surrounding terrain.

Surviving by cleverness, leopards rest and await their prey among these rocks, then spring on the unwary. Accepting their nature, neither speedy like the cheetah or strong like the lion, the leopards chose their high perches from a necessity for stealth, rather than desire for a climb. But by so doing, they had turned what could have been taken as a weakness into a strength. Proud and solitary, they are master of all they survey. It took, however, one final trip across the Serengeti before I truly understood the power of the leopard.

After a morning of scanning the infinite expanses of parched plains, we were about to eat a huge box lunch when our driver suddenly motioned for silence. Stopping our vehicle, he pointed to a ridge of low hillocks in the near distance, where a leopard, confident and alone, strode determinedly ahead. Looking to neither the right nor the left, she lifted first the legs on her left, then on her right, as she gracefully advanced. It was unusual to see a leopard on the move during the heat of the day. Perhaps after an unsuccessful night of hunting, she had only given up at daybreak to hurry back to a hidden litter of cubs. But she gave no clue as to her intent.

Charting a course as straight as a ruler, she never deviated from her route. Sometimes she disappeared behind a small, barren knoll, only to soon reappear plodding resolutely forward. Each time she was hidden from our view, our driver drove ahead of her, moving closer and closer in a kind of clandestine leapfrog. Finally, after we waited for her in a large clearing, she acknowledged our presence by halting and turning her head. She paused for a few minutes to inspect us. To my surprise, when she began walking again she chose a new course. She was heading directly toward us.

Our driver shut off the motor. As we waited, the silence was intense. What was she going to do? Did she want our food? Did she want *us* as food? How close was she going to come?

As she came nearer and nearer our driver quickly reached to roll up his window, then pulled down the elevated top of the vehicle. He seemed more than a bit nervous. Quickly following his example, I rolled up my window and then froze in amazement, not daring to saw a word.

Looking directly at me, the leopard walked straight up to my window and stared without expression at my face. She was barely three feet away! Staring back into her strangely opaque green eyes, I felt as if I were falling under a spell. Mesmerized, I suddenly knew how her prey must feel when it could no longer run away. Yet I felt no fear. Soundlessly, something passed between us. Then I blinked, and she slowly turned away.

After rounding our vehicle, she resumed her original straight, steady course, as if we had not interfered with her journey. Letting out my breath with an explosive sigh, I realized I'd been holding it for a very long time.

When evening came and we were safely back at our hotel, I couldn't forget that intense, green gaze, and knew I never would. Somehow, in some permanent way, the leopard had captured me. Yet, in some odd way I had also been her equal. Neither of us had turned and run, nor lain injured from the confrontation.

Before I came to Tanzania, disability had made me feel more vulnerable than most. However, it did not take long before this premise was in question. Watching the animals achieve their balance between vulnerability and power, what I had thought of as a weakness was revealed to be more the ineptitude of a teachable child. Perhaps I could not grow into the strength of size or tooth or speed like the cosseted baby elephants, the hidden leopard cubs, and the zebra colts that were yet to be. But I could make use of my human gift of intelligence—a gift that had already put me on wheels and even let me fly on borrowed wings—surely the most valuable strength of all.

Power and vulnerability, extended even to the extremes of life and death, will always strike a balance. Again and again, this wild land had reinforced the fact that the strong and weak, the able and disabled, are all included in the natural cycles of nature's grand plan. There was no question that I had as much a place in the world as any other. Even more importantly, I had learned that when one accepts one's own nature as completely as the leopard, whatever place one has in the world will be, like hers, a magnificent domain.

* * *

It took a camel ride, a stay in a palace, and a total eclipse of the sun in 1995 before I understood what true acceptance really means. Of course, it happened in India, that country so expert in blending differing roots to produce a common flower.

12

The Return of the Light in India: A Journey of Acceptance

As near-dawn fingers of light began to slowly uncurl into the beginning bustle on the streets of New Delhi, an elephant lumbered by our slowly moving cab. Alongside the road, a large family lay peacefully sleeping on a concrete mattress of sidewalk, warmly covered by a heavy blanket of humid summer air. The comforting familiarity of these sights made our arrival as satisfying as a long-awaited homecoming after one has been away far too long.

When two doormen and a guard smilingly hoisted my wheelchair up the steep steps of our hotel, I no longer clutched the armrests in fruitless panic, as I had on previous visits. Gladly accepting the courtesy of the lift, I rode as relaxed as a queen borne by courtiers into a favorite summer palace.

This was my introduction to a journey of acceptance. Genuine acceptance of your place in the world is an allowing of yourself to flower with firm roots in a soil of trust that things will work out. It is a soothing relaxation into circumstance when you have handed over reins of control to the gods. At some times, however, acceptance comes less easy than it does at others, such as when we later set out for an Old Delhi marketplace in a *tuk-tuk*.

As the driver of the tiny three-wheeled jitney darted rapidly in and out of the fast moving traffic, cars loomed up like springing giants, aggressively asserting that their larger size gave them right-of-way. Even the hump–backed cows, who seemed to tower above our vehicle, insolently walked directly in front of our bumper, missing it only by inches. Often, only milliseconds before impact, the driver, in true Indian fashion, determinedly injected his vehicle between lines of traffic, creating a sudden new lane with no room to spare. Trying not to cringe when it seemed that the worse was inevitable, I could only leave it in the hands of karma to decide on the outcome. When we finally disembarked at the marketplace, dodging pedestrians and livestock in the milling crowd, once an intimidating task, now seemed pleasantly tame.

"Come see my dresses," said one woman who sat in the first of a line of stalls. She motioned enthusiastically at fancifully embroidered garments pegged on a board behind her.

"No, come see my scarves, pretty lady," called a merchant farther down, waving a gilded square of silk in my direction just as a man prodded my shoulder with a tiny folding chessboard.

"You buy? You play chess?" he said, with his eyes sparkling with eagerness. "Only 600 rupees. Twenty dollars." He gave me an ingratiating smile.

"No, not interested," I said emphatically, but softened my rejection with a smile. "Good luck in finding a better customer," I added, turning slightly away.

Grinning, he ducked his head and said, "You take time to decide you want," then ran off to tap it on the shoulder of a white-suited Indian man, who gave him a dirty look and increased his pace. Together, they quickly disappeared into the busy, chattering crowd.

Except for the exotic merchandise and the colorful saris of the women who raised their voices in shrill negotiation, the scene was as familiar as my hometown street fair. But, suddenly, any sense of familiarity vanished when I noticed the cobra at the end of the first row of stalls.

In the midst of a gathering crowd, a snake charmer swayed with his flute as the deadly reptile rose up from a small wicker basket. As those watching made a space for my wheelchair in the front row, the man with the flute noticed my attention and reached to open two other wicker containers. Two more cobras oozed up to sway in rhythm with his eerie melody, mesmerized by the music and the motion. Hissing, all three snakes spread their hoods as they undulated higher and higher, their beady eyes intent on the flute.

Suddenly one struck at the snake charmer. With a hand fast as lightening, one of his friends grabbed its neck and, gripping it tightly, hauled its basket farther away before cautiously letting it go. Noting my wide-eyed attention, the flutist smiled and opened yet another basket.

I was no longer thrilled by my front-row seat when a long, thick snake—not a cobra, but one I was sure was equally deadly—oozed its way out of the last basket and headed straight toward me. The snake charmer's friends smiled and made reassuring gestures as I tried to curl my feet up high on my footrests and attempted to back away. But the crowd was now too thick and, fascinated by the spectacle, would not give way. The snake charmer, intent on his cobras, did not spare me even a moment's glance.

"T.C.," I wailed. He moved quickly to my side as the snake slithered closer. Using his bulk to force a path among the watchers, he began to wheel me away. Calling for my attention, one of the snake charmer's friends ran forward.

"You have very fine snakes," I said. "But we have to go now. Thank you."

With a tight smile, he said, "Three thousand rupees. One hundred dollars."

"Way too much," I said, surprised that he wanted a fee. Then, bowing to the inevitable, I gave him 100 rupees.

With a ferocious scowl, he almost spat as he shoved the bill back in my hands. "Just for you. All the snakes. Just for you," he said. "Three thousand rupees." His friends were putting the covers back on all but the original basket.

But the man's ploy was a usual part of the game played by street entertainers in India and I refused to be humiliated by his action. With a smile, I shrugged and began to put the money away.

"No. No!" said the man, darting forward and grabbing the money from my hand just as we broke free from the mass of watchers and headed for the street.

"I do thank you," I called back.

The smiling visage of the chess set man suddenly blocked my view of the snake man's response.

"Nice present. You want to buy? Very good present," he said hopefully.

Since my wallet was still in my hand, I offered him 300 rupees. To my surprise he dropped the set in my lap and held out his hand to take the bills.

"Thank you, Mem," he said grinning from ear to ear, then quickly disappeared back into the crowd. I smiled at my good bargaining, until I remembered that I no longer play chess.

It was almost a relief, not only to break free of the snake charmer, but to depart from noisy Delhi the next day, bound for the peace of India's desert state, Rajahstan. At the airport, however, my being in a wheelchair brought me more than the ordinary attention when I needed to use the public bathroom. A bathroom maid, one of those women who hand you a towel in return for a tip, refused to leave my side once she helped me wedge my wheelchair through the narrow outer door. To my appalled surprise, she proceeded to come right in the stall with me.

Personal functions are not always as strictly private in Third

World countries as they are at home. She appeared to believe that I needed some kind of help. No matter how hard I tried to convince her that I could handle things alone, she wouldn't leave. I soon realized that if I wanted to use the facilities I would have to be as accepting of her presence as she was of my biological needs. Once I got over the initial shock, I managed to take care of what I had come in for. Grateful that she, at least, had kept her eyes carefully averted, I made sure I tipped her well on my way out.

An American woman waiting for our flight turned to me for sympathy when the baggage clerk insisted on checking every item in her purse.

"Oh! How embarrassing," she exclaimed, blushing violently as he took everything out, including her diaphragm, and laid it out on the counter.

"It's not the worst thing that can happen," I told her, smiling. As she smilingly agreed, I thought, *If you only knew.*

After a "puddle-jumping" flight, which stopped at each of three major cities, we finally arrived in Udaipur, a place of infinite peace. Gently winding roads led us through rolling hills covered with small pale-blue houses. Finally, a large placid lake appeared directly ahead, a watery oasis nestled among the sandy-beige hills of the city. Our hotel—a former palace of the maharajah—seemed to float in the middle of the sparkling blue water like a glistening white mirage. Shaped like a confectioner's elaborate creation, its curlicued beauty seemed far more fantastical than real. It scarcely seemed that it could be a hostelry for an ordinary mortal.

When a gently rocking boat transported us to this otherworldly place, it was as if we were cradled in the arms of the gods. Up close, the palace appeared to ride on dancing rainbows where the bright water lapped softly at its base. Even when we disembarked, my wheels, mirrored in glossy marble, barely seemed to make contact with anything substantial.

Such romantic imaginings, however, soon had to give way to a more mundane reality. The smile of the bowing, mustachioed man at the desk quickly changed to a look of worry when he noticed my wheelchair. The reason was quickly apparent. I had been assigned a room on the third floor of a faerie tower, but it could only be reached by climbing three flights of stairs. After hurried consultations with a quickly called manager—tall, turbaned, and very concerned—we were quickly assigned a large suite on the main floor. The wheelchair, instead of being a liability had gotten me a room far better than I would have otherwise had. The suite came complete with an extra room for T.C., a sitting room and a balcony hanging directly over the lake.

Here I could pretend I was an Indian princess, complete with perfumed accommodations and bowing retainers. As I looked through the finely interlaced marble screen on the window, I wondered what huge kohl-rimmed dark eyes had once gazed down at the peaceful lake before me. Had she who had lived here reveled in being a cherished princess, or felt imprisoned on this isolated island? I could appreciate both emotions. Although I was treated as a cherished guest, I was as good as imprisoned in the palace.

Even if I took a boat across to the city, there was virtually no place for a wheeled visitor to go. Situated for inconvenience to enemies, this hilly community boasted nearly vertical streets that rivaled those of San Francisco. They were far too steep to allow exploration of their crowded narrow spaces from anywhere other than inside a car. Although the palace stay was like a dream come true, I was relieved when we moved on to Jodphur, where I was to trade my wheels for legs.

Ever since seeing *Lawrence of Arabia,* I had always dreamed of riding alone into the desert on a camel like a real warrior chieftain. But when I was diagnosed with MS, the dream flickered away like a sputtering film when the power wanes at the cinema. Yet the image in my mind suddenly leapt back to life when we arrived in Jodphur, and I had my long-awaited chance to ride into the dunes.

We again stayed in luxury at the palace—formal, immense, and elegant—where the current maharaja still resides. Designed for inaccessibility to enemies, this hilltop royal residence provides a real challenge to a wheelchair traveler. My room was separated from the lobby by a yawning chasm of stairs—eighteen steps down to a courtyard and eighteen steps back up. But once T.C. wrestled me across this barrier I had free run of the spacious public halls.

The imposingly formal palace was a marvelous place to explore. Multiple gleaming, marble rooms were reserved for trophies of past hunts and life-sized portraits of previous majarahas. Separate rooms set apart for smoking, drinking, and watching television were filled with even more lavish mementos of a bygone era—here an elephant-foot table, there a tiger-skin rug.

Everywhere I looked, stuffed leopards, tigers, and bears seemed to follow my every move with curiosity deep in their glassy eyes. Perhaps they suspected that I had not come only to eat and sleep like a maharajah—an Indian king. But if they knew that my dream was to be a *maharanah*—a royal desert warrior—they were in no position to tell.

Early one morning, we were driven out of town to an isolated village where two boys soon appeared with a raggedy camel. This was to be my mount. After they strapped a thick blanket on its hump, the surly beast knelt with an angry moan.

Before I could change my mind, T.C. hoisted me up, and the boys pulled my right leg over its hump. Groaning loudly, the camel suddenly lifted its rear and I slid rapidly forward. As it abruptly raised its front, I rocked just as quickly to the back.

Finding me somehow still hanging on, the camel screwed up its lips and spat.

As one of the boys pulled at the front and another used a stick from behind, the camel reluctantly moved forward. Holding tightly onto the thin material of the makeshift saddle, I looked desperately about for T.C. Seated happily in the shade with a soda, he waved a cheery good-bye.

"It's a twenty-minute ride to the dunes," he called. "We'll come with a truck when you've had enough."

I gritted my teeth and smiled. I didn't want to be a wimp and be hauled through the desert in a truck. I was here for adventure. As my mind flooded with images of *Lawrence of Arabia* and of all the former *maharanahs* dashing over endless desert dunes, I dug my heels into the camel's side and loudly yelled over my shoulder, "No thanks. No truck." And then, with as straight a back as I could manage, I rode proudly away.

Swaying and shifting with each thud of the camel's feet, I suddenly had the odd sensation of walking on four legs. As I rode here at the edge of the great Thar desert, they felt a lot more secure than my own limping two, and I soon relaxed into their rhythm. As I traveled further into the sandy, bush-dotted desert, however, the strangely pleasant sensation of walking with an extra pair of legs was gradually replaced by one far less desirable. Straddling the camel's sharp spine, I began to feel as if I were sitting on the business end of a razor blade.

Spotting my distress, the boy leading the camel stopped the beast and ran back to offer me water. Hopefully, I asked, "How much further to the dunes?"

He pointed to a distant ridge and said, smiling, "The dunes. Soon." He seemed to have little more English.

As the morning's heat intensified, we repeated this little ritual again and again. But we seemed to be getting no nearer the dunes. Trying desperately to find a comfortable position, I didn't want to give up. The desert I was seeking was not this scrub brush terrain, so much like central Arizona. My fantasy had always included endless, rolling dunes of pure sand, and I was determined to see them.

After forty long minutes we came to the ridge, but on the other side was only more flat, plant-dotted sand. With disappointment, I realized that I was nearly 500 miles east of the real desert. I wouldn't find the rolling dunes of my dreams without riding for more than a week.

Hoping against hope, I continued on for a few more minutes. But, as the now scorching sun rose higher and higher, I finally called for a halt. As soon as my groaning camel folded his legs and knelt, I slid quickly to the ground. I was so glad to be off that razor-sharp back that I didn't spare a thought for how I would get back on.

Sitting on the sand in the scant shade of a scruffy tree, I finished the bottle of water. The boys hunkered down near me. For several long minutes, we all gazed at the horizon in silence.

As the dry brown leaves above me scraped together in a tiny puff of breeze, I suddenly understood why the Rajastanis rimmed their children's eyes with kohl in hopes of intensifying their vision. As my own eyes strained to encompass the infinite horizon, I could feel my soul expanding to fill all the far spaces.

It no longer mattered that I had not found the dunes of my dreams. This was also the real desert—bleak and beige and vast. Here along the fabled Silk Road, caravans had traveled for centuries, bringing China's fine porcelains and rich silks to waiting ships on the Arabian Sea, and Arabia's famed spices to China.

Closing my eyes, I could almost hear a light tinkle of camel bells in the distance, accompanied by an ever-so-faint scent of frankincense and myrrh. When I opened them again, both boys were watching me with concern. Feeling infinite peace, I smiled at them, then asked, "How far to the truck?"

Grinning, the boys motioned back the way we'd come and said. "The truck. Soon." Within seconds, I heard the roar of the motor.

As it neared my smile broadened as I began to laugh softly to myself. It wasn't because I didn't have to get back on the camel. It was because it no longer mattered how I got back—on four feet, or two, or even on wheels. I had already gotten more than what I had come for. Not only had I come deep into the desert, but the beauty of the desert had, in some intangible way, come deep into me.

Although I hadn't found the dunes I sought, what I did find here, near the edge of the great Thar, was greater than any of the desert adventures I had dreamed. Maybe I wasn't Lawrence of Arabia or a fierce *maharanah* of Jodphur. But, to myself, I was now Kate of India, and it was more than good enough.

When we returned to the village, the women quickly gathered around me. But what attracted them was not that I had ridden one of their many village camels. As they laughingly surrounded my wheelchair, it quickly became clear that the attraction was my earrings, not only their style but their number. In each ear I wore three filigreed golden earrings bought earlier at a shop in Udiapur.

Competing for my attention, each woman brushed aside her bright veil of red, orange, or saffron-yellow silk to show me that their custom was to have two piercings in each ear. Soon they thrust forward two chuckling, naked female toddlers to demonstrate how young they were when it was done. Finally they coaxed their small boys away from a fascinated inspection of my wheelchair to show me that males had only a single piercing in one ear.

At first, I was mystified at their clucks of concern when they got around to inspecting my wrists for bracelets. Casting sympathetic glances at me when they noticed I wore none, they demonstrated the multiple gleaming white bracelets that covered their arms from shoulder to hand. Carefully calibrated in fit, ranging from quite large to very small, the bracelets were clustered so tightly it was amazing that their owners could even bend their arms.

Ultimately, frustrated by my lack of understanding, they called over our driver to translate. He explained that the bracelets were placed by the groom's family on the brides on their wedding day. Seeing me with none, the women had concluded that I was widowed, and were worried about my future. When I explained through the driver that I had never married, they appeared mystified, then patted me on the shoulder and murmured soft condolences. One sad-eyed woman came and held my hand, after asking the guide to tell me that she also could have no children. For a village woman it was a fate almost worse than death and my eyes spoke my sympathy.

Finally a very old woman spoke, while all the other women quieted in respect. Peering at me contemptuously from wise black eyes, she lengthily voiced a harsh-sounding diatribe, jabbing a blunt finger towards me several times. I shifted in my chair uncomfortably as the younger women started to smile, then covered their mouths to hide their soft laughter. I knew I was the subject of the harangue and the smiles, but couldn't understand why the old woman was so intense.

Puzzled, I turned to our driver, who finally gave me an embarrassed translation. The old woman had expressed her opinion that I must be crazy to have come to Rajahstan in a wheelchair because nothing was accessible. I had to admit she had a good point. But I smiled and told the driver to tell her that I had come because her country was so very beautiful and I had been hoping to meet such a wise woman as she. Looking unconvinced, she nodded soberly at his explanation. But then her eyes began to twinkle and she made more comments that made her audience laugh. I knew no way to explain that T.C. and I had always worked hard, all over the world, to make the most inaccessible places, accessible. There was certainly no way I could explain what it had meant to me to have ridden their camel.

* * *

The celebration of my feat had to wait until we went on to Jaipur, where the holiday of Devali was gearing up into full swing. A Hindu combination of celebrations—like New Year's Day, Christmas, Yom Kippur, and the Fourth of July all rolled into one—Devali was most of all a celebration of life.

We almost didn't get to celebrate it. As we sped through the hills around Jodphur on our way to the airport, our driver suddenly braked to a complete halt. The narrow streets were completely filled with cars and trucks, and none of them were moving. Muttering under his breath, the driver wedged our car through the tightly packed vehicles, then veered off into a field. After bumping across it at a high rate of speed he soon turned onto an almost empty dirt road. All too soon, however, he was again forced to a scowling halt. The way ahead was jammed by others who had previously had the same idea, while behind us the road was already beginning to fill. Soon, we were completely enclosed in a tight pocket of horn-blowing drivers.

Moving only a foot or two at a time on the road, we ultimately doubled back to the original street, but on the other side of the blockade. There we joined a queue of cars who, one by one, were forcing their way back onto the street. Finally it was our turn to wedge into the two-way traffic which had crammed into a space adequate only for one line of vehicles. As we edged ahead, only one more crowded intersection lay between us and the freedom of the broad airport highway.

Despite the delay caused by the detour, it appeared as though we might be able to make our plane until we suddenly came face to face with impending disaster. Near the intersection, a gasoline truck had partially run off the road and looked about ready to fall over. With its massive wheels raised up on one side by an extremely high shoulder, it was difficult to see how it could be restored to a normal position. Our driver switched off his motor and got out to join a large group of men who called instructions to others who pushed on the huge cylinder of gas from one side. Moving little more than an inch at a time, the truck driver cautiously edged his vehicle forward.

Noting several people holding cigarettes, I no longer worried about whether or not we would miss our plane. I was more concerned about the possibility that we might never be on any more planes if, in fact, the truck did go over. Suddenly there was a cheer as the truck reached an area where the high shoulder narrowed. With a thump that shook the earth, its wheels dropped back onto level road and soon, belching smoke, it was on its way.

Seeing their chance, several cars simultaneously tried to follow in its wake. At the same time the opposing traffic tried to enter the same space. Almost immediately, traffic was again at a cacophonous standstill. Our driver joined the rest in a din of honks and shouts. Noticing my wheelchair roped on the top of our car, a few men came over to offer advice. Their first concern was that my chair was coming loose on the bumpy road. After multiple hands reached out to successfully retie the rope, they joined the driver in vigorous discussion. The word "airport" was mentioned several times.

Suddenly our driver revved the engine to fever pitch, then thrust the car into gear. The random knot of men suddenly became a wedge-shaped phalanx, moving up the road like a battering ram. All other traffic gave way. Hurrying along behind our escort, there was no way we could stop when a sharp ping resounded from our undercarriage. It sounded like a flung-up stone might have hit something vital. Finally we reached the open highway, and I forgot all about it as we raced toward our destination.

When we arrived at the airport, we found the plane still waiting. So many visitors scheduled for the Jaipur flight had been caught in the same gridlock that they had rescheduled the plane for an hour later. We still had very little time to spare. As our driver parked the car, several security men ran over shouting and pointing under our wheels. Our driver got out, shrugged, then hurried back to unload my wheelchair.

The air seemed pungent with an overpowering odor of gasoline. As I sat down in my chair, I finally saw what concerned the still-agitated security crew and the growing crowd who joined them. A spreading puddle under our engine was trailed by an undulating snake of wetness where we had driven on the airport road. The ping we had heard earlier had punctured our tank. I took a deep breath when I remembered our excessive speed on the last lap of the hilly highway and thought of the overheated engine. As I let it out slowly in a long soft sigh, I accepted our safe arrival as a simple matter of karma. I could only trust that it would hold until I was a safe distance away from the car. Still, I gave extra thanks as we took off for Jaipur where our celebration of life at the Devali festival would now have special meaning.

In Jaipur, every building—from the pink multi-windowed whimsy of the Palace of the Winds to the grey, sandstone old lady, the Amber Fort—was strung with multi-colored lights for the celebration. Everywhere they could fit them in, the people had erected large imitation buildings and elaborate arches over the streets so they could display even more strings of brightly lit bulbs.

The city echoed with the rat-a-tat-tat of firecrackers, while newly built sidewalk stands groaned under the weight of huge mounds of fireworks and special candies. The babble of bargaining customers filled the air as tailors and sari shops filled to overflowing with celebrants coming to buy their holiday clothes.

Devali is the time to put on new clothes, whitewash one's house, and settle one's accounts for the year; a time to pray to the gods for favor, make peace with one's neighbors, and celebrate with dinners and gifts for the family. As elaborate fireworks light up the sky, this is also a time to joyously promenade through the streets, enjoying the lights and greeting one's friends. By day the city had the atmosphere of a bustling carnival; by night, it became an enchanted fairyland.

"Happy Devali," said the clerk at our hotel, yet another royal palace pressed into service for visitors. "Happy Devali," said the people in the streets.

"Happy Devali," I said to them all as I let myself be decorated with red ocher between my eyes. "Happy Devali," I said in the street of spices where exotic scents made my senses sing; and again in the street of tailors where red and gold silks bedazzled my eyes. I repeated the joyous greeting many times more in the street of candymakers where endless confections tempted any palate. "Happy Devali," came the echo a thousand times back. As I rolled among the crowds, I smiled and greeted them all, even the donkeys, water buffalo, and ponies whose eyes were on an exact level with mine.

But the best was still to come, made more poignant, as is usual in India, by exposure to the very worst. Soon, we hurried on to Agra to keep a rare appointment with an eclipse. After an overnight stay overlooking the Taj Mahal, we were driven at dawn to Fatehpur Sikri, a deserted city a few miles south, where we would have our best chance for an unobstructed view when the sun disappeared. But a dark shadow fell over the anticipated joy when we came across the bears.

Like great hulking Frankenstein's monsters, ominous shadowy man-shapes reared up by the side of the road each time we slowed to pass through a village. On closer inspection, they were huge, brown bears, kept close to a human master by a leash. Denied the magnificence of their wild heritage, they were a piteous sight. Chains bound their mangy paws like convicts on a chain gang. Despite extraction of all their teeth, their masters had seen fit to crudely bind their drooling jaws in rusty wire muzzles. Groaning piteously, they were kicked and clubbed until they stood upright and performed an awkward shuffling dance.

"It's illegal," said the driver, when I protested in tears. "Those men are gypsies," he said with a scowl. "The government has tried to put them in jail but they take their animals and hide. The bears look sad, yes. For them life is hard. But their masters are very, very poor."

Having noticed the gypsies were almost as tattered and thin as their bears, I knew he accepted what he saw because of what he knew as truth. But, speechless with distress, I knew that, with my own brand of truth, I could never find their actions justifiable.

Mercifully, such issues were quickly banished as we joined the happy melee of eclipse watchers at the ruins of Fatehpur Sikri, a city that died centuries ago when it ran out of water. Frequently honking his horn, our driver interwove our car among hundreds of holidaying pedestrians heading up a steep dirt road. We made it almost to the huge doors of the broken-walled palace before a guard insisted that the car be parked.

Rooftop watchers already crowded the top of the palace, but T.C. found me a spot on the less congested grassy area at its base. While he wandered off to investigate all the "experts" who had spread out about the hills with giant special cameras and fanciful electronic equipment, I chatted with members of a large boisterous group seated nearby. Complete with backpacks and blankets, they had been settled in place since the previous night.

"Where are you from?" I asked, then listened in fascination.

"I'm from England; she's from Sweden," said a bearded young man with his arm around a girl with long blond hair.

"California," said another who sported a deep tan.

"Pakistan," was the soft response of a young, graceful woman with a small child clinging to her sari.

"Ontario," "Australia," "China," "South Africa," sang out several voices. No one even glanced at my wheelchair. Our shared interest in this special event made for a mutual acceptance which gave no importance to such a small detail.

"Don't use your video camera until the eclipse is total," advised the Canadian. "Otherwise its circuits will burn out."

"It's safe to film it with your regular camera but, unless you have a heavy filter, you won't get a good picture," allowed the Australian.

"It's getting cold," said the Chinese girl, and shivered as if to prove her point.

And sure enough, the morning sky was darkening as a cold wind began to pull itself out of the very earth. Stars began to appear even though it was already past eight o'clock in the morning. Sneaking quick, cautious peeps because of the glare, I couldn't see that the sun had changed until the Englishman loaned me his special Mylar glasses. Despite the intensity of its light, only a sliver of sun could be seen when viewed through this special filter.

Suddenly dogs in the valley started barking wildly and a peacock high atop the castle shrieked in outrage as the sun, bringer of light and life, totally vanished. All that was left was a thin ring of dull fire. Through the odd darkness—neither twilight nor night—a roar bounced off the hills from all directions as hundreds of throats cried out with awe and excitement. A primal yell was torn from my own throat as tears of joy suddenly welled in my eyes.

At that moment, I understood why, not all that long ago, villagers had locked themselves and their animals in their houses and prayed to the gods when a total eclipse of the sun occurred. In their primitive belief, the sun god had expressed anger with them when he took away their light. It was time to pray and to hope things would come out right. Shivering in the sudden cold, I even understood why they threw out all their food and clothing, fearing contamination from the unknown forces that had stolen their sun. Despite my more modern views, it was a fearsome moment, and my neck prickled as its fine hairs raised up in atavistic respect.

After a minute, which seemed more like an hour, the sun and moon began to put themselves right as they fortunately always do. At first, "Bailey's beads" appeared as the craters of the moon became outlined against a first splinter of reappearing sun. Then, even though I had been waiting for it, I didn't realize I was seeing the next rare celestial event—the "Diamond Ring"— before the sudden dazzle nearly blinded me. As the moon moved away, a small, globular piece of sun escaped its shadow and suddenly appeared intermingled with the fiery corona. It did indeed look like an engagement ring sparkling with the bright light of promise. When I was able to open my stinging eyes again, the sun was now too bright to look at, even though the skies were still as dark as the inside of a thundercloud.

When we returned to Agra I felt a sense of victory. The light had gone away and come back again and all was right with the world. In some way I knew right then that this age-old story was also the story of my life. It was indeed a happy Devali, a time to acknowledge new beginnings and accept the return of the light.

* * *

On the Frankfurt–Washington, D.C., lap of the long journey back to the United States, I began talking with the passengers near me on the plane.

"Where have you come from?" I asked. "Have you had a holiday in Germany or did you connect from somewhere else?"

"Germany," they said, and I smiled.

"I've come from seeing the solar eclipse in India," I said. Then, to my surprise, I found myself adding proudly, "I travel in a wheelchair, you know."

As they smiled and nodded I knew that only I understood how very far I had come since that first tentative venture into China when I had shrunk from being photographed in a wheelchair. Now, instead of thinking of it as an albatross, I knew it had allowed me to fly like an eagle, carrying me high and proud into the far horizons of my dreams. It had been a long, hard journey to acceptance but, with the sun flashing off my wings, I knew I had finally found my way home.

Epilogue

When I finished this book, I was not the same person who began it. I had meant simply to write tales about my travels to wondrous lands and tell of the grand adventures I'd had. Instead, the story turned out to be about my coming of age as, traveling around the world, I simultaneously took internal journeys to more unexpected destinations.

Most people only come of age twice, once at puberty and again as they enter their senior years. By becoming disabled in the prime of life, I had an extra chance to get it right. As I confronted issues about disability and even some concerning death itself, the title, *Disappearing Windmills,* came from the discovery that, like Don Quixote, I had at first misinterpreted what I was dealing with.

Almost all the issues with which I was confronted appeared at first, as giant, clanking monsters, formidable and frightening, which had to be challenged with sword and lance if life was to be rewarding. But, one by one, they were revealed to be mere windmills, quite normal features on the landscape of life, which, if understood, could be used as a perpetual source of personal power. Once they were properly identified, they soon disappeared as a matter of concern.

Although my concerns about the initial issues have essentially disappeared there are real battles that must be fought. This time the "monsters" are far more dangerous. Unfortunately, they are also, in reality, quite ordinary but they are far more difficult to dismiss as a matter of concern. Rarely possible to make disappear entirely, they are prejudice, discrimination, and intolerance—all fathered by lack of understanding—and I do not have to journey far from home to find them.

Even in our "land of the free," the supposedly inalienable rights of all people are not universally allowed to those who happen to be disabled. Lack of adequate access, both physical and attitudinal, can sometimes effectively interrupt our pursuit of happiness. Lack of attention to the development of adequate assistive devices can also compromise our liberty. Instead of dealing with us as we are, our society, as expressed through the research it funds, seems more interested in making our disabilities disappear.

Although some of the disabled may wish for this as much as for life itself, others understand it's not essential for their fulfillment. The cure-versus-care controversy is an old one among the disabled, if asked what they want. But no matter where one stands, it can be seen that some efforts to wipe out disability, if taken to the furthest extreme, could backfire and threaten our very right to life itself.

Since a "cure" has not been found for many disabling conditions, at least with the minimal amount our society has allotted for the necessary research, scientists are being encouraged to search for causative genes. If used properly such knowledge might benefit us all. But it could also allow doctors to play God and change the very substance of our lives, or enable mothers to choose not to let us even be born. It is truly frightening to envision a world where not only disabilities but the disabled themselves could be made not to exist.

Some segments of our society already feel that merely having a disability is justification for a Kevorkianesque permanent solution. Unfortunately, even some that are disabled buy into this myth and end their lives long before their time. "Poor thing, she's probably better off," will say one left behind. "I understand his choice. If I were in his position, I'd do away with myself too," echos another. But whose choice was it really?

But no matter how much our autonomy may be threatened by able-bodied people who presume to speak for us, and our human rights ignored by those who don't understand us, we won't disappear. Most of us don't want to. I don't want to.

In America, probably because of proclaimed ideals of freedom and equality, we are slowly realizing more physical access to our society. Instead of being hidden away in homes and institutions, we are out and about in a diverse variety of settings. But efforts to gain attitudinal access are still mired in the mud. When living and visible in our society, we are, unfortunately, a reminder of possibilities, a reminder that anyone at anytime can join our club. Because of lack of knowledge and understanding, our fraternity is deemed far more undesirable than it actually is.

Now that I've come of age, I know it is time that we be recognized and treated as an integral part of society and not merely of a subset termed "disabled." We are not all the same, and each of us can supply valuable and diverse points of view. We are not even all agreed on "disability issues," such as whether we want money allocated to search for "cures" or used to create more assistive devices and opportunities for noninstitutional personal care. But we all deserve to be heard on any issue that affects us.

Unfortunately, as we attempt to take our rightful place in our own society, we discover ourselves to be in many ways disenfranchised. We must battle a strong cultural bias that often does not recognize our voices as valid even in matters that directly concern us. Such a societal blind spot allows, for example, an able-bodied person to state a bathroom is "in compliance" with the ADA (Americans with Disabilities Act), although a person in a wheelchair can see in an instant that, even if he can get in, the facilities are positioned in a way that makes it impossible to get close enough to use them.

As in the case of my "helper" in the Forbidden City, those

physically more able too often act for the disabled without asking what we want; and listen only grudgingly to our point of view if we give them an answer anyway. But we have to keep trying, if we expect the "monsters" of prejudice, intolerance, and discrimination against the disabled to be driven off and ultimately disappear.

I have found far too few willing to acknowledge even the possibility that disabled people can live as happily and successfully as anyone else when left to their own devices—although I'm living proof of it. Contrary to popular supposition among those who are able, I do not consider myself lacking because I can't match the physical achievements of others. Instead, I rejoice in being the very best disabled person I can be.

When it comes right down to it, many of my grand adventures, while traveling through the world and through life, have come *because* I am disabled and not *in spite* of it. I can honestly say, though many might doubt, I'm disabled and *proud of it!* In the face of that battle cry, I've seen even the monsters of societal bias hesitate, then begin to run away.

When I declare my happiness, I'm sometimes told that I am an exception. But I wasn't created to be an "inspiration" as some misguidedly attempt to label me. I was created, like any other of God's children, to be happy and live in love. And I am, and I do. As another traveler said, after he also mistook windmills for monsters, "Fortune disposes our affairs better than we ourselves could have desired." Reflecting on my life and its "disappearing windmills," I couldn't agree more.

Acknowledgments

This book and the journeys described in it would not have been possible without Howard McCoy III of Accessible Journeys, who, never allowing me to think of myself as unable, arranged the trips and was the real T.C. (my Trusty Companion). He also was worth his weight in gold as a devil's advocate through his critical reading of my manuscript and insistence that I could write better.

I also owe a great debt to Leslie Komarnicki, an English teacher at the time, who believed in me and told me I could write, even as she corrected my many mistakes.

In addition I need to thank Heather Long, who patiently read all the chapter drafts and encouraged me to keep on with my writing when I sometimes became discouraged.

A final thank you to Will DeRuve, Publisher of *ACCESS to Travel* magazine, whose publishing of brief stories extracted from my book helped me believe that the general public might like to read it.

My heartfelt thank-you goes to all of them, and to each I say it a thousand times. And to each I say "I love you," but I'm sure they already know how I feel.